Problems in
Mathematical Analysis

MONOGRAPHS AND TEXTBOOKS IN
PURE AND APPLIED MATHEMATICS

67. *J. K. Beem and P. E. Ehrlich,* Global Lorentzian Geometry (1981)
68. *D. L. Armacost,* The Structure of Locally Compact Abelian Groups (1981)
69. *J. W. Brewer and M. K. Smith, eds.,* Emmy Noether: A Tribute to Her Life and Work (1981)
70. *K. H. Kim,* Boolean Matrix Theory and Applications (1982)
71. *T. W. Wieting,* The Mathematical Theory of Chromatic Plane Ornaments (1982)
72. *D. B. Gauld,* Differential Topology: An Introduction (1982)
73. *R. L. Faber,* Foundations of Euclidean and Non-Euclidean Geometry (1983)
74. *M. Carmeli,* Statistical Theory and Random Matrices (1983)
75. *J. H. Carruth, J. A. Hildebrant, and R. J. Koch,* The Theory of Topological Semigroups (1983)
76. *R. L. Faber,* Differential Geometry and Relativity Theory: An Introduction (1983)
77. *S. Barnett,* Polynomials and Linear Control Systems (1983)
78. *G. Karpilovsky,* Commutative Group Algebras (1983)
79. *F. Van Oystaeyen and A. Verschoren,* Relative Invariants of Rings: The Commutative Theory (1983)
80. *I. Vaisman,* A First Course in Differential Geometry (1984)
81. *G. W. Swan,* Applications of Optimal Control Theory in Biomedicine (1984)
82. *T. Petrie and J. D. Randall,* Transformation Groups on Manifolds (1984)
83. *K. Goebel and S. Reich,* Uniform Convexity, Hyperbolic Geometry, and Nonexpansive Mappings (1984)
84. *T. Albu and C. Năstăsescu,* Relative Finiteness in Module Theory (1984)
85. *K. Hrbacek and T. Jech,* Introduction to Set Theory, Second Edition, Revised and Expanded (1984)
86. *F. Van Oystaeyen and A. Verschoren,* Relative Invariants of Rings: The Noncommutative Theory (1984)
87. *B. R. McDonald,* Linear Algebra Over Commutative Rings (1984)
88. *M. Namba,* Geometry of Projective Algebraic Curves (1984)
89. *G. F. Webb,* Theory of Nonlinear Age-Dependent Population Dynamics (1985)
90. *M. R. Bremner, R. V. Moody, and J. Patera,* Tables of Dominant Weight Multiplicities for Representations of Simple Lie Algebras (1985)
91. *A. E. Fekete,* Real Linear Algebra (1985)
92. *S. B. Chae,* Holomorphy and Calculus in Normed Spaces (1985)
93. *A. J. Jerri,* Introduction to Integral Equations with Applications (1985)
94. *G. Karpilovsky,* Projective Representations of Finite Groups (1985)
95. *L. Narici and E. Beckenstein,* Topological Vector Spaces (1985)
96. *J. Weeks,* The Shape of Space: How to Visualize Surfaces and Three-Dimensional Manifolds (1985)
97. *P. R. Gribik and K. O. Kortanek,* Extremal Methods of Operations Research (1985)
98. *J.-A. Chao and W. A. Woyczynski, eds.,* Probability Theory and Harmonic Analysis (1986)
99. *G. D. Crown, M. H. Fenrick, and R. J. Valenza,* Abstract Algebra (1986)
100. *J. H. Carruth, J. A. Hildebrant, and R. J. Koch,* The Theory of Topological Semigroups, Volume 2 (1986)

101. *R. S. Doran and V. A. Belfi*, Characterizations of C*-Algebras: The Gelfand-Naimark Theorems (1986)

102. *M. W. Jeter*, Mathematical Programming: An Introduction to Optimization (1986)

103. *M. Altman*, A Unified Theory of Nonlinear Operator and Evolution Equations with Applications: A New Approach to Nonlinear Partial Differential Equations (1986)

104. *A. Verschoren*, Relative Invariants of Sheaves (1987)

105. *R. A. Usmani*, Applied Linear Algebra (1987)

106. *P. Blass and J. Lang*, Zariski Surfaces and Differential Equations in Characteristic p > 0 (1987)

107. *J. A. Reneke, R. E. Fennell, and R. B. Minton.* Structured Hereditary Systems (1987)

108. *H. Busemann and B. B. Phadke*, Spaces with Distinguished Geodesics (1987)

109. *R. Harte*, Invertibility and Singularity for Bounded Linear Operators (1988).

110. *G. S. Ladde, V. Lakshmikantham, and B. G. Zhang*, Oscillation Theory of Differential Equations with Deviating Arguments (1987)

111. *L. Dudkin, I. Rabinovich, and I. Vakhutinsky*, Iterative Aggregation Theory: Mathematical Methods of Coordinating Detailed and Aggregate Problems in Large Control Systems (1987)

112. *T. Okubo*, Differential Geometry (1987)

113. *D. L. Stancl and M. L. Stancl*, Real Analysis with Point-Set Topology (1987)

114. *T. C. Gard*, Introduction to Stochastic Differential Equations (1988)

115. *S. S. Abhyankar*, Enumerative Combinatorics of Young Tableaux (1988)

116. *H. Strade and R. Farnsteiner*, Modular Lie Algebras and Their Representations (1988)

117. *J. A. Huckaba*, Commutative Rings with Zero Divisors (1988)

118. *W. D. Wallis*, Combinatorial Designs (1988)

119. *W. Więsław*, Topological Fields (1988)

120. *G. Karpilovsky*, Field Theory: Classical Foundations and Multiplicative Groups (1988)

121. *S. Caenepeel and F. Van Oystaeyen*, Brauer Groups and the Cohomology of Graded Rings (1989)

122. *W. Kozlowski*, Modular Function Spaces (1988)

123. *E. Lowen-Colebunders*, Function Classes of Cauchy Continuous Maps (1989)

124. *M. Pavel*, Fundamentals of Pattern Recognition (1989)

125. *V. Lakshmikantham, S. Leela, and A. A. Martynyuk*, Stability Analysis of Nonlinear Systems (1989)

126. *R. Sivaramakrishnan*, The Classical Theory of Arithmetic Functions (1989)

127. *N. A. Watson*, Parabolic Equations on an Infinite Strip (1989)

128. *K. J. Hastings*, Introduction to the Mathematics of Operations Research (1989)

129. *B. Fine*, Algebraic Theory of the Bianchi Groups (1989)

130. *D. N. Dikranjan, I. R. Prodanov, and L. N. Stoyanov*, Topological Groups: Characters, Dualities, and Minimal Group Topologies (1989)

Other Volumes in Preparation

Problems in Mathematical Analysis

Piotr Biler
Alfred Witkowski
University of Wrocław
Wrocław, Poland

CRC Press
Taylor & Francis Group
Boca Raton London New York

CRC Press is an imprint of the
Taylor & Francis Group, an **informa** business

CRC Press
Taylor & Francis Group
6000 Broken Sound Parkway NW, Suite 300
Boca Raton, FL 33487-2742

First issued in paperback 2019

© 1990 by Taylor Francis Group, LLC
CRC Press is an imprint of Taylor & Francis Group, an Informa business

No claim to original U.S. Government works

ISBN-13: 978-0-8247-8312-9 (hbk)
ISBN-13: 978-0-367-40322-5 (pbk)

Library of Congress catalog number- 89-25685

Library of Congress Cataloging-in-Publication Data

Catalog record is available from the Library of Congress

Visit the Taylor & Francis Web site at
http://www.taylorandfrancis.com

and the CRC Press Web site at
http://www.crcpress.com

Preface

This volume offers an unusual collection of problems in mathematical analysis. First, most are neither standard nor easy. Second, readers will find here both the classical topics (so classical that they are even forgotten!) and the modern ones. Third, there are also original problems on an advanced level that may lead readers to their own research projects. Some of the problems suggest new proofs of well-known theorems. They are given here to attract the readers' attention to the original and/or elegant method of proof.

The references are divided into two parts. The first contains classical problem books and well-known textbooks or monographs. There is little overlap between the contents of these volumes and this volume empty (we have avoided reproducing here even the most beautiful pearls from Halmos [H], the *AMM Problem Book* [OD], Pólya-Stegö [P-S], Rudin [R1], [R2], and Demidovič [D]). The second part of the references only includes essentially the cited sources. Some of them rarely serve as problem books and some are not distributed widely.

For the readers' convenience we give, whenever possible, English language references to a problem in particular to the *American Mathematical Monthly*, even if the problem has appeared earlier elsewhere or in another language.

Besides not restricting ourselves to the first published form of a problem, we have, in some cases preferred to transform slightly the formulation in order to obtain the problem as commonly referred to in the literature. Thus the note "for example" in a reference indicates that we have cited merely one of several valid sources.

The answers and the hints (or rather suggestions) are in general extremely laconic because there is no unique way to the result. Active readers will find their own methods and would not have this pleasure if the full reasoning were given (moreover, this would make the volume enormous).

The brevity of references does not necessarily mean that either an exercise is easy or it is our original problem. The place of a problem in the chapter does not indicate either its difficulty level or the necessary knowledge of the theory useful in the solving process.

It is our pleasure to thank all our friends and colleagues from mathematical departments of various institutions in Wrocław and Paris for their interest, discussions, and helpful remarks. The warmest thanks are due to Czesław Ryll-Nardzewski and Ryszard Szwarc for original problems; to Jean-Marie Strelcyn, who encouraged us to prepare the English translation of the original Polish version; to Françoise Piquard and Hervé Queffélec, who offered us invaluable comments; to Alano Ancona, Marek Bożejko, Alain Damlamian, Hubert Delange, Pierre Goetgheluck, Bolesław Kopociński, Zbigniew Lipecki, Jean-François Méla, Wojciech Mydlarczyk, Tadeusz Pytlik, Luc Robbiano, Zbigniew Skoczylas, Stanisław J. Szarek, and Krzysztof Tabisz for various remarks and encouragement, and especially to Bohdan Mincer for his constant editor's assistance.

Good luck in problem solving, since there is no better way to approach mathematics.

Piotr Biler
Alfred Witkowski

Contents

Problems in
Mathematical Analysis

1

Real and Complex Numbers

Topics covered include:
prime numbers, representation of reals as series, group and topological structure of real numbers, polynomials, inequalities, rational and irrational numbers, symmetric functions, geometric properties of complex numbers, various problems.

1.1 Show that an irrational power of an irrational number can be rational.

1.2 Prove that if $c > 8/3$, then there exists a real number θ such that $[\theta^{c^n}]$ is prime for every positive integer n.

1.3 Show that there exists a real number θ such that each number of the

form $\begin{bmatrix} 2^\theta \\ \cdots \\ 2^2 \end{bmatrix}$ is prime.

1.4 Let (a_n) be an arbitrary sequence of integers greater than 1. Prove that every real number $x \in [0, 1)$ can be represented as $x = \sum_{k=1}^{\infty} x_k/a_1 a_2 \cdots a_k$, where $x_k \in \{0, 1, \ldots, a_k - 1\}$. Give necessary and sufficient conditions for the existence of two such representations of the same number x.

1.5 Show that every $x \in (0, 1]$ can be represented as $x = \sum_{k=1}^{\infty} 1/n_k$, where (n_k) is a sequence of positive integers such that $n_{k+1}/n_k \in \{2, 3, 4\}$.

1.6 Prove that if $a_n \neq 0$, $n = 1, 2, \ldots$ and $\lim_{n \to \infty} a_n = 0$, then for every real number x there exists the integer sequences (k_n), (m_n) such that $x = \sum_{n=1}^{\infty} k_n a_n$ and $x = \prod_{n=1}^{\infty} m_n a_n$.

1.7 Show that if for every natural n $0 < a_n < \sum_{k=n+1}^{\infty} a_k$ and $\sum_{n=1}^{\infty} a_n = 1$, then for each $x \in (0, 1)$ there exists a subsequence (k_p) satisfying $\sum_{p=1}^{\infty} a_{k_p} = x$.

1.8 Given a countable subset C of $(0, 1)$ find necessary and sufficient conditions for the existence for every $x \in (0, 1)$ of a rearrangement of C in a sequence (c_k) such that $\sum_{k=1}^{\infty} c_k / 2^k = x$.

1.9 Consider expansions of real numbers $x \in (0, 1)$ $x = \sum_{n=1}^{\infty} x_n / r^n$ with the fixed natural number $r > 1$ and $x_n \in \{0, 1, \ldots, r - 1\}$. Prove that for almost all x this expansion is normal in the sense of Borel, that is, the frequency of each digit $0, 1, \ldots, r - 1$ in the sequence (x_n) equals $1/r$.

1.10 Show that if $(\sqrt{5} - 1)/2 = \sum_{k=1}^{\infty} 2^{-n_k}$, where n_k are positive integers, then $n_k \leqslant 5 \cdot 2^{k-2} - 1$.

1.11 Define for $x = \sum_{n=1}^{\infty} x_n / p^n$, $x_n \in \{0, 1, \ldots, p - 1\}$, $f(x) = \sum_{n=1}^{\infty} x_n / q^n$, where $1 < p \leqslant q$ are integers. Compute $\int_0^1 f(x)\, dx$.

1.12 Consider a permutation σ of the set of digits $\{0, \ldots, 9\}$ and define the function $f : [0, 1) \to [0, 1]$ by the formula: $f(x) = \sum_{n=1}^{\infty} \sigma(x_n)/10^n$ if $x = \sum_{n=1}^{\infty} x_n / 10^n$ is the decimal expansion (which does not contain almost all 9s). Study continuity and differentiability properties of f. Compute $\int_0^1 f(x)\, dx$.

1.13 Determine the topological structure and the Lebesgue measure of the set of reals in $[0, 1]$ whose infinite decimal expansions contain all the digits $1, 2, \ldots, 9$.

1.14 Prove that for a strictly increasing natural sequence (n_k) satisfying $\lim_{n \to \infty} n_k^{1/2^k} = \infty$, $\sum_{k=1}^{\infty} 1/n_k$ is irrational.

1.15 Consider two natural sequences (p_n), (q_n) such that $p_n/q_n(q_n - 1) \geqslant p_{n+1}/(q_{n+1} - 1)$ and $x = \sum_{n=1}^{\infty} p_n/q_n < \infty$. Prove that x is irrational if and only if the inequality in the condition above is strict infinitely many times.

1.16 Show that for every natural sequence $1 < a_1 < a_2 < \cdots$, $\sum_{n=1}^{\infty} 2^{a_n}/a_n!$ is irrational.

1.17 Prove that for every integer a, $|a| > 1$, $\prod_{n=1}^{\infty} (1 - a^{-n})$ is irrational.

1.18 Show that for infinitely many rationals $\theta \in (0, 1)$ $g(\theta) = \sum_{n=1}^{\infty} \theta/n(n + \theta)$ is irrational.

1.19 Prove that given the natural numbers k and m, the number $S(m, k) = \sum_{n=1}^{\infty} (n(mn + k))^{-1}$ is rational if and only if m divides k.

1.20 Show that $\sum_{n=1}^{N}(\{x + n/N\} - \frac{1}{2}) = \{Nx\} - \frac{1}{2}$ for every natural N, where $\{x\}$ denotes the fractional part of x.

1.21 Show that $\int_{0}^{1}(\{ax\} - \frac{1}{2})(\{bx\} - \frac{1}{2}) dx = (12ab)^{-1}$ for the relatively prime natural numbers a and b.

1.22 Compute $\int_{0}^{1} S_n(x) dx$, where $S_n(x) = \sum_{k=1}^{n}(\{a_k x\} - \frac{1}{2})$ and $a_1, \ldots, a_n$ are positive integers. Show that $\int_{0}^{1} S_n^2(x) dx = (12)^{-1} \sum_{1 \leq k, m \leq n} (a_k, a_m)/[a_k, a_m]$.

1.23 Find $\max_{\sigma \in S_n} \sum_{k=1}^{n} |\sigma(k) - k|$, where S_n is the symmetric group of permutations of n symbols.

1.24 Prove that every finite sequence of $n^2 + 1$ reals contains a monotone subsequence with $n + 1$ terms.

1.25 Suppose that a strictly increasing natural sequence (a_n) satisfies $\limsup_{n \to \infty} n^{-1} \sum_{a_k \leq n} 1 = 1$. Show that (a_n) contains a subsequence of mutually relatively prime numbers.

1.26 Determine all natural numbers that are sums of consecutive integers.

1.27 Let $f_n(x) = \min\{|x - m/n| : m \in \mathbb{Z}\}$, $n = 1, 2, \ldots$. For which real values of x does the series $\sum_{n=1}^{\infty} f_n(x)$ converge? Answer the same question for $f_n(x) = \min\{|x - p/q| : p, q \in \mathbb{Z}, |q| \leq n\}$.

1.28 Consider the first digits of the consecutive powers of 2: $1, 2, 4, 8, 1, 3, \ldots$. Does this sequence contain 7? Which digit is more frequent in this sequence: 7 or 8?

1.29 Let (r_n) be an enumeration of all rationals. Show that $\mathbb{R} \backslash \bigcup_{n=1}^{\infty} (r_n - n^{-2}, r_n + n^{-2}) \neq \emptyset$. Does there exist an enumeration of rationals (s_n) such that $\mathbb{R} \backslash \bigcup_{n=1}^{\infty} (s_n - 1/n, s_n + 1/n) \neq \emptyset$?

1.30 Show that there exists an enumeration (x_n) of the rationals in $(0, 1)$ such that $\sum_{n=1}^{\infty} \prod_{i=1}^{n} x_i$ diverges.

1.31 Let (x_n) be a sequence of positive numbers which does not tend to zero. Prove that the set $A = \{a : x_n \to 0 \text{ in } \mathbb{R}/a\mathbb{Z}\}$ is of the Lebesgue measure zero.

1.32 Does there exist a subset $S \subseteq \mathbb{R}$ such that $\inf\{a > 0 : S + a = \mathbb{R} \backslash S\} = 0$?

1.33 Does there exist a bounded sequence of mutually different reals whose every translation has only finite intersection with the Cantor ternary set?

1.34 Let $f_0(x) = \cosh x$, $f_1(x) = \sinh x$. Show that for $r = 0, 1$

$$f_r(x_1 + \cdots + x_n) = \sum_{\substack{s_i = 0, 1 \\ s_1 + \cdots + s_n \equiv r \pmod 2}} \prod_{i=1}^{n} f_{s_i}(x_i)$$

1.35 Prove that the multiplicative group $\mathbb{C}^*$ of nonzero complex numbers is isomorphic to the group of complex numbers of modulus 1.

1.36 Show that for every proper subgroup G of the (additive) group $\mathbb{Q}$, $\mathbb{Q}/G$ is infinite.

1.37 Does there exist a Hamel basis of $\mathbb{R}$ over $\mathbb{Q}$ which is closed under multiplication?

1.38 Prove that every nondiscrete locally compact topological group whose every proper closed subgroup is discrete is topologically equivalent to $\mathbb{R}$ or $\mathbb{T}$.

1.39 Show that there does not exist any countable partition of reals into closed sets.

1.40 Let $S \subseteq \mathbb{Z}$ be a cofinite subset. Show that there exists two infinite sets S_1 and S_2 such that $S_1 + S_2 = S$ and every $s \in S$ has a unique representation $s = s_1 + s_2$ with $s_i \in S_i$.

1.41 For a given real x consider the set of all 0-1 sequences (ε_n) such that $\sum_{n=1}^{\infty} \varepsilon_n/n = x$. Determine the cardinality of this set.

1.42 Show that for natural numbers $0 < a_1 < \cdots < a_n$ $\Pi_{1 \leqslant i \leqslant j \leqslant n}(x^{a_i} - x^{a_j})/$ $(x^i - x^j)$ is a polynomial.

1.43 Find the remainder when $(x + a)^n$ is divided by $(x + b)^m$.

1.44 Show that a real monic polynomial vanishing at $x = 1$, $x = 2$ has a coefficient less than -2.

1.45 Show that every polynomial p with rational coefficients such that $p^{-1}(\mathbb{Q}) \subseteq \mathbb{Q}$ must be linear.

1.46 Show that every complex polynomial p such that $p(\mathbb{R}) \subseteq \mathbb{R}$ and $p(\mathbb{C} \setminus \mathbb{R}) \subseteq \mathbb{C} \setminus \mathbb{R}$ must be linear.

1.47 The polynomials $P(x) = x^2 + ax + b$ and $Q(x) = x^2 + px + q$ have a common root. Find a quadratic polynomial whose roots coincide with the remaining roots of P and Q.

1.48 Show that if two monic polynomials commute with a polynomial $a(x) \not\equiv x$, then they commute.

1.49 Show that every polynomial has a multiple which is a polynomial of $x^{1000000}$.

1.50 Compute $\sum_{j=1}^{n-1} (1 - \exp(2\pi ij/n))^{-1}$.

1.51 Let $f(x) = (x - x_1) \cdots (x - x_n)$, $x_i \neq x_j$ for $i \neq j$, $g(x) = x^{n-1} + a_{n-2}x^{n-2} + \cdots + a_0$. Show that $\sum_{j=1}^{n} g(x_j)/f'(x_j) = 1$.

1.52 Let $a_1 \leqslant \cdots \leqslant a_n$ be the roots of a polynomial of degree n and $b_1 \leqslant \cdots \leqslant b_{n-1}$ be the roots of its derivative. Compute $\Sigma_{i,j}(b_i - a_j)^{-1}$.

1.53 Prove the Laguerre theorem: If $a_1 < \cdots < a_n$ denotes the roots of a polynomial of degree n and $b_1 < \cdots < b_{n-1}$ the roots of its derivative, then b_j is not contained in the small interval contiguous to a_j obtained by dividing (a_j, a_{j+1}) into n equal intervals.

1.54 Show that for every $n = 0, 1, 2, \ldots$ there exists the unique polynomial $B_n(x)$ satisfying the identity $\int_x^{x+1} B_n(t)\, dt = x^n$. Determine $B_n(x)$. The numbers $B_n = B_n(0)$ are called the Bernoulli numbers. Show that $B_{2k+1} > 0$ and $B_{2k} B_{2k+1} < 0$ for $k > 0$. Find a formula for the sum $s_n(m) = 1^n + 2^n + \cdots + m^n$.

1.55 Prove that for arbitrary real $a_1, \ldots, a_n$, $(\Sigma_{k=1}^n a_k/k)^2 \leqslant \Sigma_{k=1}^n \Sigma_{j=1}^n a_k a_j/(k+j-1)$.

1.56 Show that for nonnegative $b_1, \ldots, b_n$ such that $b_1 + \cdots + b_n = b$, $\Sigma_{k=1}^{n-1} b_k b_{k+1} \leqslant b^2/4$.

1.57 Let $a_1, \ldots, a_n$ be the lengths of the sides of a convex n-gon of area K. Show that $\Sigma_{k=1}^n a_k^2 \geqslant 4K \tan(\pi/n)$.

1.58 Suppose that for positive numbers $p_1, \ldots, p_n$ and $q_1, \ldots, q_n$ $\Sigma_{k=1}^n p_k = \Sigma_{k=1}^n q_k$. Prove the inequality (important in the information theory) $\Sigma_{k=1}^n p_k \log p_k \geqslant \Sigma_{k=1}^n p_k \log q_k$.

1.59 Show that for any permutation $b_1, \ldots, b_n$ of positive numbers $a_1, \ldots, a_n$, $\Sigma_{k=1}^n a_k/b_k \geqslant n$. For which permutation does this sum attain its maximum?

1.60 Let $a_1, \ldots, a_n$ be positive numbers and $s_k = s_k(a_1, \ldots, a_n)$, $k = 1, \ldots, n$, be the kth symmetric polynomial, $b_k = s_k/\binom{n}{k}$. Prove the inequalities:

(i) $b_1 \geqslant b_2^{1/2}$
(ii) $b_k^2 \geqslant b_{k-1} b_{k+1}$ the Newton inequality
(iii) $b_k^{1/k} \geqslant b_{k+1}^{1/(k+1)}$ the Maclaurin inequality

1.61 Show that if for all natural n $[na] + [nb] = [nc] + [nd]$ then at least one of the numbers $a+b$, $a-c$, $a-d$ is an integer.

1.62 Prove the identity $[x] + [x + 1/n] + \cdots + [x + (n-1)/n] = [nx]$ for all real x and natural n.

1.63 Show the inequality $[nx] \geqslant [x] + [2x]/2 + \cdots + [nx]/n$ for all real x and natural n.

1.64 If $q = (\sqrt{5} + 1)/2$, then for every natural n $[q[qn]] + 1 = [q^2 n]$. Does there exist another q with the same property?

1.65 Show that if $x > 1$ is irrational, then the sets $\{[nx] : n \in \mathbb{N}\}$ and $\{[nx/(x - 1)] : n \in \mathbb{N}\}$ form a partition of $\mathbb{N}$.

1.66 Let (u_n) be the sequence 1, 2, 4, 5, 7, 9, 10, 12, 14, 16, 17, ... where one odd number is followed by two evens, then three odds, etc. Show that $u_n = 2n - [(1 + (8n - 7)^{1/2})/2]$.

1.67 Find the accumulation points of the set $\{\sqrt{n} - \sqrt{m} : m, n \in \mathbb{N}\}$.

1.68 Compute $\limsup_{n \to \infty} |a^n - b^n|^{1/n}$ for $a \neq b \in \mathbb{C}$.

1.69 The image of every straight line in the complex plane under the transformation $z \mapsto z^2$ is a parabola. Find its vertex.

1.70 Prove that there is no complex function such that its rth iterate is equal to $az^2 + bz + c$, $a \neq 0$, $r \geq 2$.

1.71 Prove that for arbitrary polynomials P, Q and constant k $P(D)[e^{kx}Q(x)]|_{x=0} = Q(D)[P(x)]|_{x=k}$, where $D = d/dx$.

1.72 Let D be the differentiation operator d/dx on the space of polynomials of degree $\leq n$. Find a simpler expression for $T = I + D + D^2/2! + \cdots + D^n/n!$.

1.73 Determine maximal multiplicity of the roots of the equation $P(x) = x(x - 1) \cdots (x - n) = \lambda$.

1.74 Show that if $f(z) = z^n + a_1 z^{n-1} + \cdots + a_n$ is a real polynomial, $a_n = \pm 1$, then the product of the roots of f lying outside the unit disk is less than $(1 + a_1^2 + \cdots + a_n^2)^{1/2}$.

1.75 Show that all roots of the equation $nz^n = 1 + z + \cdots + z^n$ lie in the unit disk.

1.76 Let λ be a root of the polynomial $f(z) = z^n + a_1 z^{n-1} + \cdots + a_n$, where $1 \geq a_1 \geq a_2 \geq \cdots \geq a_n > 0$. Show that if $|\lambda| \geq 1$, then λ is a root of unity.

1.77 Define the real polynomials f and g by the identity $(1 + ix)^m = f(x) + ig(x)$. Prove that for every real a and b, all roots of the polynomial $af + bg$ are real.

1.78 Show that the polynomial $1 + x + x^2/2! + \cdots + x^{2n}/(2n)!$ has no real roots and the polynomial $1 + x + x^2/2! + \cdots + x^n/n!$ has no multiple roots.

1.79 Determine all natural n such that all nonzero roots of the equation $(1 + z)^n = 1 + z^n$ lie on the unit circle.

1.80 Determine $M_n = \min_{p \in P_n} (\max_{|z|=1} |p(z)|)$, where P_n is the set of complex polynomials of the form $z^n + a_1 z^{n-1} + \cdots + a_{n-1} z + 1$.

1.81 Compute $\min_{z\in\mathbb{C}}\max\{|1 + z|, |1 + z^2|\}$.

1.82 Calculate $\min |2 - (z_1 + z_2)(z_3 + z_4) + z_1 z_2 z_3 z_4|$, where z_j are complex numbers of modulus greater than $r \geqslant 2$.

1.83 Show that for every complex a the equation $e^z = (a + z)/(a - z)$ has no solution in the quadrant $\{z : \operatorname{Re} z > 0, \operatorname{Im} z > 0\}$.

1.84 Show that for arbitrary complex numbers $z_1, \ldots, z_n$ there exists a subset B of indexes $\{1, \ldots, n\}$ such that $|\Sigma_{k \in B} z_k| \geqslant \pi^{-1} \Sigma_{k=1}^{n} |z_k|$.

1.85 Determine the number of real roots of the equation $(x + 1)^x + \cdots + (x + n)^x = (x + n + 1)^x$, $n \geqslant 2$.

1.86 Prove that for every natural k $\Sigma_{0 \leqslant x_i \leqslant n} \min(x_1, \ldots, x_k) = \Sigma_{m=0}^{n} m^k$, $x_i \in \mathbb{N}$.

1.87 Discuss the positive solutions of the system

$$\begin{cases} x^{x+y} = y^a \\ y^{x+y} = x^b \end{cases}$$

with real parameters a and b.

1.88 Find all real solutions of the system

$$\begin{cases} x + y + z = w \\ \dfrac{1}{x} + \dfrac{1}{y} + \dfrac{1}{z} = \dfrac{1}{w} \end{cases}$$

1.89 Prove that if $3 | k \in \mathbb{Z}$, then the equation $x + 1/x + y + 1/y = k$ has no positive rational solutions.

1.90 Find all solutions of the system

$$\begin{cases} y^3 - 9x^2 + 27x - 27 = 0 \\ z^3 - 9y^2 + 27y - 27 = 0 \\ x^3 - 9z^2 + 27z - 27 = 0 \end{cases}$$

1.91 Determine all real c such that the equation $\sqrt{cx}/(x^2 - 1) = 1/(x - 1)$ has exactly one solution.

1.92 For $0 < a_1 \leqslant \cdots \leqslant a_n < 1$ define the function

$$f(t) = \sum_{k=1}^{n} \left(\frac{2}{\pi} \arctan a_k t - \frac{k}{n+1} \right)^2$$

Prove that for $t > 0$, $f(t) < f(0)$.

1.93 Prove that the real sequence (a_n) defined by $a_{n+1} = |a_n| - a_{n-1}$ is periodic.

1.94 For real $c_1, \ldots, c_n$ and $k \geqslant 1$ define

$$c_{n+k} = 1 - \cfrac{c_{n+k-1}}{1 - \cfrac{c_{n+k-2}}{1 - \cfrac{}{\ddots \cfrac{c_{k+1}}{1 - c_k}}}}$$

Prove that for all $k \geqslant 1$ $c_{n+k+3} = c_k$.

1.95 Show that for $a > 0$

$$\sum_{\substack{p,q=1 \\ (p,q)=1}}^{\infty} (a^{p+q} - 1)^{-1} = (a-1)^{-2}.$$

1.96 Let $\mathscr{F}$ denote the family of all finite subsets of $\mathbb{N}$. Given $p \in (0,1)$ and $n \in \mathbb{N}$, construct a function $\varphi: 2^{\mathbb{N}} \to \mathbb{R}$ of the form $\varphi(X) = \sum_{Y \in \mathscr{F}} a_Y(-1)^{|X \cap Y|}$, $a_Y \geqslant 0$, $\sum_{Y \in \mathscr{F}} a_Y = 1$, such that if $\min X \leqslant n$, then $\varphi(X) = p$.

1.97 Let (D_n) be a sequence of disjoint disks of radii r_n contained in the unit disk. Show that if $\sum_{n=1}^{\infty} r_n < \infty$, then $\sum_{n=1}^{\infty} r_n^2 < 1$.

1.98 Prove that there exists a square containing the sequence of disjoint squares of sides $1, \frac{1}{2}, \frac{1}{3}, \ldots$.

1.99 (K_n) is a sequence of squares of area a_n. Show that if $\sum_{n=1}^{\infty} a_n = \infty$, then the squares (K_n) can cover the plane.

1.100 Let C_0 be a circle of radius 1 centered at $(-1, 1)$, C_1 is a circle of radius 1 centered at $(1, 1)$ and for $k \geqslant 2$ C_k is the circle tangent to C_0, C_{k-1} and the Ox axis. Calculate the sum of the areas of the disks bounded by C_1, $C_2, \ldots$.

1.101 The regular n-gon $A_1 \cdots A_n$ rolls without slipping along the line ℓ. At time $t = 0$ the side $A_1 A_2$ is on ℓ. Let B_i denote the point where the vertex A_1 is while $A_i A_{i+1}$ lies on ℓ. Show that
   (i) The area of the set bounded by polygonal line $B_1 \cdots B_n$ and ℓ equals to three areas of the polygon $A_1 \cdots A_n$.
   (ii) The angles in the vertices $B_2, \ldots, B_{n-1}$ are equal.
   (iii) The length of the polygonal line $B_1 \cdots B_n$ equals $4(R + r)$, where R and r are the radii of the circles circumscribed on and inscribed in $A_1 \cdots A_n$.
Compare these results with the well-known properties of the cycloid.

1.102 Let L_n, A_n denote the length of the curve $r^n = a^n \sin n\theta$, $0 \leqslant \theta \leqslant 2\pi$ and the area of the region bounded by this curve. Compute $\lim_{n \to \infty} A_n$ and $\lim_{n \to \infty} L_n/n$.

1.103 2^n balls of radius $1/2$ centered at $(\pm 1/2, \ldots, \pm 1/2)$ are inscribed in the n-dimensional cube $[-1,1]^n$. Compute the diameter of the small ball centered at $(0, \ldots, 0)$ and tangent to all of these balls.

1.104 The parabolas $y = x^2$ and $y = -x^2$ are tangent. The upper parabola rolls without slipping along the lower one. Find the position of the moving foci.

1.105 Prove that a transformation of the plane which preserves all rational distances is an isometry.

1.106 Consider a transformation of n-dimensional Euclidean space that preserves the distance 1. Is it an isometry?

1.107 Let $f : \mathbb{Z} \to \mathbb{Z}$ be an injection which preserves the relation $|i - j| \leqslant k$ for a fixed integer k. Show that f is an isometry.

1.108 Let E be a compact metric space and $f : E \to E$ is such that $d(f(x), f(y)) \geqslant d(x, y)$. Show that f is an isometry.

1.109 Given natural n, construct a partition of the plane into n congruent connected sets.

1.110 Is it possible to represent $\mathbb{R}^3$ as a disjoint union of mutually not parallel lines?

1.111 Let G be a finite group of affine transformations of $\mathbb{R}^n$. Show that there exists a common fixed point of the group G.

1.112 How many triangles can be constructed using the segments of length $1, 2, \ldots, n$?

1.113 A point inside a triangle is located at the distances 2, $\sqrt{2}$, $\sqrt{3} - 1$ from the vertices. Determine the maximal possible area of this triangle.

1.114 The triangle with vertices $(0,0)$, $(0,p)$, $(p,0)$, where p is a prime number, is divided into p triangles by the segments joining $(0,0)$ with $(p-r,r)$, $r = 1, \ldots, p - 1$. Show that every one of $p - 2$ middle triangles contains the same number of lattice points.

1.115 A rectangle P is the union of disjoint rectangles P_j, $j = 1, \ldots, n$. The length of at least one side of each P_j is rational. Prove that P has the same property.

1.116 R is a convex polygon, $d(x, y)$ denotes the distance from the point (x, y) to the interior of R. Show that there exist the constants a, b, and c

independent of R such that $\int_R \int_R e^{-d(x,y)} \, dxdy = a + bL + cA$, where L is the perimeter of R and A is its area.

1.117 Let C be the (smooth) boundary of a bounded convex set and T is the triangle of maximal perimeter inscribed in C. Prove that the normal to C in the vertex of T coincides with the bisectrix of T. Is the converse true?

1.118 If a segment AB of length 1 is rotated about the fixed point B of angle π to the final position BA', then the length of the trace of the point A equals π. Let us allow B to move also. Determine the minimal sum of lengths of the traces of A and B necessary to move the segment AB to the position BA'.

1.119 Farmer Jones has a cart with square wheels. However it suits his needs since with it he is able to travel the washboard road without any bumping. Assuming no slipping of the wheels describe the washboard road.

1.120 Let there be a series of cups of equal capacity full of water and arranged one below another. Pour into the first cup an equal quantity of wine at constant rate and let the overflow in each cup go into the cup just below it. Assuming that complete mixture takes place instantaneously, find the amount of wine in each cup at any time t and at the end of the process at time T.

1.121 Let $w_1 = 0$, $w_2 = 1$, $w_{n+2} = w_n w_{n+1}$ for $n \geqslant 1$ as a word in $\{0,1\}$ alphabet and finally the infinite word S is $w_1 w_2 w_3 \ldots$. Show that the nth symbol in S is equal to $[n\gamma] - [(n-1)\gamma]$, where $\gamma = (\sqrt{5} - 1)/2$.

1.122 Consider the set $A_n = \{(\varepsilon_1, \ldots, \varepsilon_n) : \varepsilon_i = 0, 1\}$ and the maps $\ell, \imath : A_n \to A_{n-1}$ defined by $\ell(\varepsilon_1, \ldots, \varepsilon_n) = (\varepsilon_2, \ldots, \varepsilon_n)$, $\imath(\varepsilon_1, \ldots, \varepsilon_n) = (\varepsilon_1, \ldots, \varepsilon_{n-1})$. Show that for every subset B of A_n containing at least 2^{n-1} elements, $\ell(B) \cap \imath(B) \neq \varnothing$. Estimate the number of elements in this intersection.

1.123 There are some filling stations along the circuit and the total amount of gasoline in these stations suffices to make a circuit trip by car. Show that there exists a station such that starting from it, this trip can be made.

1.124 A submarine was spotted from a torpedo boat. After that the submarine submerged and went straight away at constant speed v_1 in an unknown direction. The speed, v_2, of the torpedo boat is greater than v_1 and the destroyer should appear just over the submarine in order to destroy it. Determine the strategy of pursuit.

1.125 Find the ellipse enclosing the minimal area containing three given points which are not collinear.

1.126 Is it possible to divide a square into an odd number of triangles of equal area?

1.127 Is the curve $\{(0, 1 + \cos t, 1 + \sin t):|t| \leqslant \pi\}$ contractible in $\mathbb{R}^3 \setminus \{(\cos s, \sin s, 1/s):s > 0\}$?

1.128 Given a convex region R in $\mathbb{R}^2$ of area 1, find the minimal number d such that the diameter of the image of R by an area preserving affine transformation is not greater than d.

1.129 Construct a dense set of points with rational coordinates in $\mathbb{R}^2$ and irrational mutual distances.

1.130 Does there exist a point in $\mathbb{R}^2$ with rational distances to the points $(0,0)$, $(0, 1)$, $(1,0)$, $(1, 1)$?

1.131 Consider the sequence defined by $a_1 = 1$, $a_{n+1} = (1 + a_1^2 + \cdots + a_n^2)/n$. Are all a_n integers?

1.132 Given n points $a_1, \ldots, a_n$ in $\mathbb{R}^3$, $n \geqslant 2$, show that there exists a unique ball in $\mathbb{R}^3$ such that
   (i) All a_j lie in this ball.
   (ii) The center of the ball is a convex combination of a_j lying on the sphere.

1.133 Compute the probability that two natural numbers are relatively prime.

1.134 Show that the area of each section of a tetrahedron is less than the area of its maximal face.

2

Sequences

Topics covered include:
limits, recursive sequences, functional sequences.

2.1 Find the limits $\lim_{n \to \infty} (2k^{1/n} - 1)^n$, $\lim_{n \to \infty} (2n^{1/n} - 1)^n/n^2$.

2.2 Calculate the radius of convergence of the power series $\sum_{n=1}^{\infty} a_n x^n$, where $a_n = \int_0^n \exp(t^2/n) \, dt$.

2.3 Find the limit $\lim_{n \to \infty} \int_0^{\sqrt{n}} (1 - x^2/n)^n \, dx$.

2.4 Show that the sequence $a_n = (1 + 1/n)^{n^2} n! \, n^{-(n+\frac{1}{2})}$ decreases and find its limit.

2.5 Show that for $f(n) = \sum_{i=1}^{n} \sum_{j=1}^{n} \binom{n}{i}\binom{n}{j} i^{n-j} j^{n-i} \lim_{n \to \infty} n^{-1}(f(n))^{1/2n}$
$\log n = 1/e$.

2.6 Study the asymptotic behavior of the sequence $a_n = \sum_{\substack{j,k>1 \\ j+k=n}} (jk)^{-1}$.

2.7 Let (a_n) and (b_n) be positive sequences such that $\lim_{n \to \infty} a_n^n = a$, $\lim_{n \to \infty} b_n^n = b$, $a, b \in (0, \infty)$, and suppose that positive numbers p and q satisfy $p + q = 1$. Compute $\lim_{n \to \infty} (pa_n + qb_n)^n$.

2.8   Given real number $x \in (0, 1)$ let $f_k(x)$ denote the number of multiples mx in the interval $[k, k + 1)$, $k, m = 1, 2, \ldots$. Prove the convergence of the sequence $(f_1(x) \cdots f_n(x))^{1/n}$ and determine its limit.

2.9   Let $S(x) = \lim_{n \to \infty} 2\pi^{-1} \arctan nx$. Compute $\lim_{m \to \infty} S(\sin^2(\pi m! x))$.

2.10   Show that $(\frac{x+1}{x-1})^{1/2} = \Pi_{n=1}^{\infty} (1 + 1/q_n)$, where $q_1 = x > 1$, $q_{n+1} = 2q_n^2 - 1$.

2.11   Show that the sequence $s_n = \Sigma_{k=n}^{pn} 1/k$, $n = 1, 2, \ldots$, converges for every $p > 1$. Consider the sequence $t_n = \Sigma_{k=n}^{pn} f(1/k)$ for a function f in C^1. Determine $\lim_{n \to \infty} t_n$ for $f(x) = \log(1 + x)$.

2.12   Show that there exists the limit u of the sequence $u_n = \Sigma_{k=1}^{n} (k^2 + n^2)^{-1/2}$. Estimate the rate of convergence of $u_n - u$.

2.13   Compute $\lim_{n \to \infty} \Sigma_{k=1}^{n^2} n/(n^2 + k^2)$.

2.14   Find $\lim_{n \to \infty} n^{-1} \Sigma_{k=1}^{n} ([2n/k] - 2[n/k])$.

2.15   Determine $\lim_{n \to \infty} \Pi_{k=1}^{n} (1 + k/n^2)$.

2.16   Calculate $\lim_{n \to \infty} \Pi_{k=1}^{n} (1 + k^2/n^3)$.

2.17   Determine $\lim_{n \to \infty} (\log \log n)^{-1} \Sigma_{k=2}^{n} (k \log k)^{-1}$.

2.18   For $a > 0$ calculate $\lim_{n \to \infty} \log^a n \Sigma_{k=n+1}^{\infty} (k \log^{1+a} k)^{-1}$.

2.19   Study the existence of the limit of $\Sigma_{k=1}^{n-1} (n^{-1} \tan k\pi/2n)^p$.

2.20   Given a natural number p, compute $\lim_{n \to \infty} (p^{-1} \Sigma_{k=1}^{p} k^{p/n})^n$.

2.21   Find $\lim_{n \to \infty} (n + 1)^{-ns} \Sigma_{k=1}^{n} k^{ns}$.

2.22   Calculate $\lim_{n \to \infty} (\Pi_{k=1}^{n} (kn)^{1/(n+k)})^{1/\log n}$.

2.23   Study the sequence $u_n = \Sigma_{k=1}^{n} (n + k)^{-a}(n + k + 1)^{-b}$ for $a, b > 0$.

2.24   Study the convergence of the sequences $x_n = \sqrt{1! \sqrt{2! \ldots \sqrt{n!}}}$, $y_n = \sqrt{2\sqrt{3 \ldots \sqrt{n}}}$.

2.25   Study the asymptotics of the expression $\Sigma_{k=1}^{n-1} \log(n - k)/k - \log^2 n$.

2.26   Let $s_n = 1 + \frac{n-1}{n+2} + \frac{n-1}{n+2} \cdot \frac{n-2}{n+3} + \cdots + \frac{n-1}{n+2} \cdot \frac{n-2}{n+3} \cdots \frac{1}{2n}$, $n = 1, 2, \ldots$. Show that $\lim_{n \to \infty} s_n/n^{1/2} = \pi^{1/2}/2$.

2.27   Suppose that (a_n) is a real sequence such that $\lim_{n \to \infty} n^x a_n = a$ for some real x. Show that $\lim_{n \to \infty} n^x (a_1 \cdots a_n)^{1/n} = ae^x$.

2.28   Prove that $\lim_{n \to \infty} (1 + 1/n)(1 + 2/n)^{1/2} \cdots (1 + n/n)^{1/n} = e^{\pi^2/12}$.

2.29 (a_n) is an arithmetic progression with positive terms. Compute $\lim_{n \to \infty} n(a_1 \cdots a_n)^{1/n}/(a_1 + \cdots + a_n)$.

2.30 Let A_n, G_n denote the arithmetic and the geometric means of $\binom{n}{0}$, $\binom{n}{1}, \ldots, \binom{n}{n}$. Show that $\lim_{n \to \infty} A_n^{1/n} = 2$, $\lim_{n \to \infty} G_n^{1/n} = e^{1/2}$.

2.31 Given real numbers $a_1, \ldots, a_k$ compute $\lim_{n \to \infty} (((n - a_1) \cdots (n - a_k))^{1/k} - n)$.

2.32 Given two positive numbers a_1 and a_2, take alternately the arithmetic and geometric means: $a_3 = (a_1 + a_2)/2$, $a_4 = \sqrt{a_2 a_3}$ etc. Find $\lim_{n \to \infty} a_n$.

2.33 Show that $I(a, b) = \int_0^{\pi/2} (a^2 \cos^2 t + b^2 \sin^2 t)^{-1/2} dt = \pi/2l$, where l is the arithmetic-geometric mean of positive numbers a and b, that is, the common limit of the sequences (a_n), (b_n) defined by $a_0 = a$, $b_0 = b$, $a_{n+1} = \sqrt{a_n b_n}$, $b_{n+1} = (a_n + b_n)/2$.

2.34 Find $\liminf_{n \to \infty} |\sin n|^{1/n}$.

2.35 Give an explicit formula for the general term of the sequence (p_n) $p_1 = 1$, $p_{n+1} = \sum_{k=1}^{n} p_k p_{n+1-k}$.

2.36 Define (x_n) by induction $x_2 = 1$, $x_3 = 2$, $x_n = (n - 1)(x_{n-1} + x_{n-2})$. Determine x_n.

2.37 Show that if $0 < y_0 < 1/c$ and $y_{n+1} = y_n(2 - cy_n)$, then $\lim_{n \to \infty} y_n = 1/c$.

2.38 Study convergence of the sequences $a_n = \sqrt{1 + \sqrt{1 + \cdots + \sqrt{1}}}$, $b_n = \sqrt{1 + \sqrt{2 + \cdots + \sqrt{n}}}$.

2.39 Compute $\sqrt{1 + 2\sqrt{1 + 3\sqrt{1 + 4\sqrt{\cdots}}}}$.

2.40 Determine $\lim_{n \to \infty} s_n$ for the recursive sequence $s_1 = \sqrt{2}$, $s_{n+1} = (2 + s_n^{\frac{1}{2}})^{\frac{1}{2}}$.

2.41 Let $a_1, b_1 \in (0, 1)$ and $a_{n+1} = a_1(1 - a_n - b_n) + a_n$, $b_{n+1} = b_1(1 - a_n - b_n) + b_n$. Show that the sequences (a_n) and (b_n) converge and compute their limits.

2.42 Find the limit of the recursive sequence $x_1 = \sqrt{2}$, $x_{n+1} = 2^{x_n/2}$.

2.43 Let x be a positive real number. Determine the limit of the sequence $x_1 = (xx^b)^a$, $x_{n+1} = (x(xx_n)^b)^a$.

2.44 Let $a_0 = 0$, $a_{n+1} = 1 + \sin(a_n - 1)$. Calculate $\lim_{n \to \infty} \sum_{k=1}^{n} a_k/n$.

2.45 Construct a sequence (A_n) of points on an interval such that $A_{n+2} \in A_n A_{n+1}$ and $|A_n A_{n+2}| \cdot |A_{n+1} A_n| = |A_{n+2} A_{n+1}|^2$. Show that the sequence $|A_1 A_n|$ converges and compute its limit.

2.46 Let (p_n) be a dense set in the interval $(0, 1)$ and let a_n and b_n denote the lengths of the intervals obtained by division by p_n of one of the intervals determined by the points $0, 1, p_1, \ldots, p_{n-1}$. Compute $\sum_{n=1}^{\infty} a_n b_n (a_n + b_n)$.

2.47 Find the limit of the sequence (a_n) defined as $a_0 = 0$, $a_1 = 1/2$, $a_{n+1} = (1 + a_n + a_{n-1}^3)/3$.

2.48 Prove that if $w_1 \geqslant w_2 \geqslant \cdots > 0$, $\sum_{n=1}^{\infty} w_n = \infty$ and (a_n) satisfy the condition $(1 + w_n)a_n = -(a_{n-2} + a_{n+2})$, then $\lim_{n \to \infty} a_n = 0$.

2.49 Show that for every $t > 0$ the sequence defined by $x_{n+1} = x_n - x_n^2/n^{1+t}$, $x_0 = 1$, has the limit $f(t)$ and $f(t) = t + O(t^2)$ as $t \to 0^+$.

2.50 Suppose that the sequence (x_n) satisfies
  (i) $x_n + ax_{n-1} + x_{n-2} = 0$.
  (ii) x_n tends to ∞.
  (iii) $x_n \neq 0$.
  Show that $\sum_{n=0}^{\infty} (x_n x_{n+1})^{-1} = (x_0(x_1 - \lambda x_0))^{-1}$, where λ is the root of the equation $\lambda^2 + a\lambda + 1 = 0$ with $|\lambda| < 1$.

2.51 Determine the limit $\lim_{n \to \infty} 2^n(2 - a_n)^{\frac{1}{2}}$, where $a_0 = 0$, $a_{n+1} = (a_n + 2)^{1/2}$.

2.52 Study the behavior of the sequence $(u_n^{1/n})$, where $u_0 \in (0, 1)$ and $u_{n+1} = (1 - (1 - u_n)^{1/2})/2$.

2.53 Compute $\prod_{n=1}^{\infty} c_n$, where $c_0 > -1$ and $c_{n+1} = ((1 + c_n)/2)^{1/2}$. Supposing additionally that $|c_0| < 1$ determine the limits of the sequences $a_n = 4^n(1 - c_n)$, $b_n = 4^n(\prod_{k=1}^{\infty} c_k - \prod_{k=1}^{n} c_k)$.

2.54 Study the convergence of the sequence (z_n), where $z_0 \in \mathbb{C}$, $z_{n+1} = (z_n + c)^{1/2}$ and roots are taken with positive imaginary parts.

2.55 Construct a complex sequence $(z_n)_{n \in \mathbb{Z}}$ such that $z_n \neq 0$, $z_{n+1} = z_n^2 + z_n$ and $\lim_{n \to -\infty} z_n = \lim_{n \to \infty} z_n = 0$.

2.56 Show that if a_0 and a_1 are positive and $a_{n+1} = a_n^{1/2} + a_{n-1}^{1/2}$, then the sequence (a_n) converges.

2.57 For which values of a_1 does the recursive sequence $a_{n+1} = a_n^2 - 2$ converge?

2.58 Define the sequence (a_n) as follows: $a_1 = x$, $a_{n+1} = (2 + 1/n)a_n - 1$. Show that (a_n) converges for a unique value of x. Determine this value and the limit of (a_n).

2.59 Study the convergence of the sequence (x_n), where $x_0 = 0$, $x_{n+1} = x_n - c(x_n^2 - a)$ for different values of parameters a and c.

2.60 Prove that if $x_0^m \leqslant D < (x_0 + 1)^m$, then the limit of the recursive sequence $x_{n+1} = x_n + (D - x_n^m)/m(x_n + 1)^{m-1}$ equals $D^{1/m}$ (the Hobson formula).

2.61 Describe all possible limits of the sequence (u_n), where $u_{n+1} = (4n + 2)u_n + u_{n-1}$ for different values of u_0, u_1. Consider in particular $u_0 = 0, u_1 = 1$; $u_0 = 1, u_1 = 0$. Give necessary and sufficient conditions on u_0 and u_1 for convergence of the sequence (u_n).

2.62 Define for $0 < y_0 < x_0 < 1$, $r > 0$, $s \geqslant 1$ $x_{n+1} = y_n + x_n^r(x_n - y_n)^s$, $y_{n+1} = y_n + y_n^r(x_n - y_n)^s$. Show that $\lim_{n \to \infty} x_n = \lim_{n \to \infty} y_n$.

2.63 Prove that if a_0 and a_1 are real numbers and $a_{n+1} = a_n + 2a_{n-1}/(n + 1)$, then the sequence (a_n/n^2) converges. Find its limit.

2.64 Show that if $a_0 = 1$ and $a_n = a_{[n/2]} + a_{[n/3]} + a_{[n/6]}$, then $\lim_{n \to \infty} a_n/n = 12/\log 432$.

2.65 Construct a sequence (a_k) such that $\Sigma_{k=0}^{\infty} 2^{nk}a_k = (-1)^n$ and $\Sigma_{k=0}^{\infty} 2^{nk}|a_k| < \infty$ for every n (the Seeley lemma).

2.66 Determine the limits of the sequences $b_{n+1} = \int_0^1 \min(x, a_n) \, dx$, $a_{n+1} = \int_0^1 \max(x, b_n) \, dx$.

2.67 Compute the limits of the sequences $a_{n+1} = \int_0^1 \min(x, b_n, c_n) \, dx$, $b_{n+1} = \int_0^1 \mathrm{mid}(x, a_n, c_n) \, dx$, $c_{n+1} = \int_0^1 \max(x, a_n, b_n) \, dx$.

2.68 Show that if $c_1 \in (0, 1)$ and $c_{n+1} = c_n(1 - c_n)$, then $\lim_{n \to \infty} n(1 - nc_n)/\log n = 1$.

2.69 Let $x_1 = 1$, $x_{n+1} = x_n + 1/\Sigma_{k=1}^n x_k$. Show that $x_n \simeq \sqrt{2 \log n}$.

2.70 Consider the recursive sequence $x_{n+1} = \sin x_n$ with $x_1 \in (0, \pi)$. Show that $x_n \simeq \sqrt{3/n}$.

2.71 It is easy to see that the recursive sequence $a_{n+1} = a_n(1 - a_n^2)$ with $a_1 \in (0, 1)$ tends to zero. Estimate the rate of convergence.

2.72 Show that for every real sequence (a_n) the following conditions are equivalent:
(i) $\lim_{n \to \infty} \Sigma_{k=1}^n e^{ia_k}/n = t$.
(ii) $\lim_{n \to \infty} \Sigma_{k=1}^{n^2} e^{ia_k}/n^2 = t$.

2.73 Suppose that the sequences (u_n) and (a_n) satisfy the conditions: $u_n \geqslant u_{n-1} - a_{n-1}$, $a_n \geqslant 0$, $\Sigma_{n=1}^{\infty} a_n < \infty$, $u_n \leqslant M < \infty$. Show that the sequence (u_n) converges.

2.74 Show that if $u_n > 0$ and the limit $\lim_{n \to \infty} n(1 - u_{n+1}/u_n)$ exists (maybe infinite), then the limit $\lim_{n \to \infty} \log(1/u_n)/\log n$ also exists.

2.75 Show that if $\lim_{n \to \infty} (a_n - a_{n-2}) = 0$, then $\lim_{n \to \infty} (a_n - a_{n-1})/n = 0$.

2.76 Prove that for every positive sequence (a_n) $\lim \sup_{n \to \infty} n((1 + a_{n+1})/a_n - 1) \geqslant 1$. Show that 1 on the right hand side is the best possible constant.

2.77 Prove that for any positive sequence (a_n) $\lim \sup_{n \to \infty} ((a_1 + a_{n+1})/a_n)^n \geqslant e$.

2.78 Given a positive sequence (t_n) define the sequence (a_n) by $a_1 = 1$, $a_{n+1} = a_n + t_n a_n^{-1}$. Prove that the sequence (a_n) converges if and only if the series $\Sigma_{n=1}^{\infty} t_n$ also converges.

2.79 Consider two complex sequences (a_n) and (b_n). Suppose that (b_n) converges and $b_n = a_n - t a_{n+1}$, where t is a fixed complex number. Show that if $|t| > 1$ or $|t| < 1$ and $\lim_{n \to \infty} a_n t^n = 0$, then the sequence (a_n) converges.

2.80 Prove that if the series $\Sigma_{n=1}^{\infty} u_n$ converges and the sequence $(u_n - u_{n+1})$ decreases, then $\lim_{n \to \infty} (u_{n+1}^{-1} - u_n^{-1}) = \infty$.

2.81 Show that if (t_n) is a positive sequence such that $\Sigma_{n=1}^{\infty} t_n = 1$, then $\lim_{N \to \infty} N^{-1} \Sigma_{n=1}^{\infty} \{N t_n\} = 0$ ($\{x\} = x - [x]$).

2.82 Prove the Tauberian theorem of Littlewood: If (a_n) is a complex sequence such that for every n $|n a_n| \leqslant 1$ and $\lim_{x \to 1} \Sigma_{n=1}^{\infty} a_n x^n = 0$, then $\lim_{n \to \infty} (a_0 + \cdots + a_n) = 0$.

2.83 A real sequence (k_n) satisfies the following conditions: $k_1 > 1$, $k_1 + \cdots + k_n < 2k_n$ for $n = 2, 3, \ldots$. Show that there exists $q > 1$ such that $k_n > q^n$ for all n.

2.84 Suppose that for a real sequence (a_n) $a_m + a_n - 1 < a_{m+n} < a_m + a_n + 1$. Prove that there exists a finite limit $\lim_{n \to \infty} a_n/n = w$ and $nw - 1 < a_n < nw + 1$.

2.85 A nondecreasing positive sequence (a_n) satisfies the inequality $a_{mn} \geqslant m a_n$ for all m and n and $\sup_n a_n/n < \infty$. Does the sequence (a_n/n) converge?

2.86 For a real sequence (a_n) let $s_n = a_0 + \cdots + a_n$. Suppose that for some constant C and all n $\Sigma_{k=1}^{n} k|a_k| \leqslant Cn$, $\lim_{n \to \infty} (s_0 + \cdots + s_n)/(n+1) = s$. Prove that $\lim_{n \to \infty} (s_0^2 + \cdots + s_n^2)/(n+1) = s^2$.

2.87 Prove that if the real sequence (a_n) satisfies the condition $\lim_{n \to \infty} a_n \Sigma_{j=1}^{n} a_j^2 = 1$, then $\lim_{n \to \infty} (3n)^{1/3} a_n = 1$.

2.88 (a_n) is a nonnegative sequence, $s_n = \Sigma_{k=0}^{n} a_k$. Show that if $\lim_{n \to \infty} s_n = \infty$, $\lim_{n \to \infty} a_n/s_n = 0$, then the convergence radius of the series $\Sigma_{n=0}^{\infty} a_n x^n$ equals 1.

2.89 Suppose that $\lim_{(x,y)\to(0,0)} F(x, y) = 0$ and a sequence (a_n) satisfies
(i) $0 < a_{n+m} < F(a_n, a_m)$, n, $m \in \mathbb{N}$.
(ii) $\lim_{n\to\infty} (a_1 + \cdots + a_n)/n = 0$.
Show that $\lim_{n\to\infty} a_n = 0$.

2.90 Determine conditions imposed on the sequences (a_n) and (b_n), which guarantee the existence of two constants A and B, $AB \neq 0$, such that the sequence $(Aa_n + Bb_n)$ converges.

2.91 Given two reals a and b, let $F(a, b)$ denote the sequence $([na + b])_{n\in\mathbb{N}}$. For which pairs (a, b) does $F(x, y) = F(a, b)$ imply $(x, y) = (a, b)$?

2.92 Show that if for reals $a_1, \ldots, a_k$ $\lim_{n\to\infty} \sin(na_1) \cdots \sin(na_k) = 0$, then at least one of the a_j is a multiple of π.

2.93 Let (u_n) strictly increase to infinity and $\lim_{n\to\infty} u_{n+1}/u_n = 1$. Show that $\sum_{k=1}^{n} (u_k - u_{k-1})/u_k \simeq \log u_n$.

2.94 Show that if a series with terms converging to zero is Cesàro summable to 0, then there exists a subsequence of the sequence of partial sums of this series converging to 0.

2.95 Prove that given a bounded sequence (a_n), $a_n \neq 0$, there exists a subsequence (b_n) such that (b_{n+1}/b_n) converges.

2.96 Define for a real sequence (x_n) and a positive sequence (p_n) such that $\sum_{n=1}^{\infty} p_n = \infty$ new sequence $y_n = \sum_{k=1}^{n} p_k x_k / \sum_{k=1}^{n} p_k$. Show that $\lim\inf_{n\to\infty} x_n \leqslant \lim\inf_{n\to\infty} y_n \leqslant \lim\sup_{n\to\infty} y_n \leqslant \lim\sup_{n\to\infty} x_n$. Moreover, (p_n) can be chosen so that the set of limit points of (y_n) contains all limit points of (x_n).

2.97 Let $C(s)$ denote the set of all limits of convergent subsequences of a bounded sequence (s_n). Show that
(i) If $(s_n - s_{n-1})$ converges to zero, then $C(s)$ is connected, but not conversely.
(ii) $C(s)$ is connected if and only if there exists a subsequence (y_n) of (s_n) such that $C(s) = C(y)$ and $(y_n - y_{n-1})$ converges to zero.

2.98 Suppose that for a real sequence (a_n) the subsequences (a_{pn}) are increasing and bounded from above for every prime p. Show that the sequence (a_n) with all terms a_p deleted converges. What can be said about (a_n) if (a_{pn}) are convergent for all prime p?

2.99 Assume that the sequence (a_n) in a locally compact Hausdorff space has the property: for every convergent subsequence (a_{n_k}), the subsequence (a_{n_k+1}) converges to the same limit. Show that either (a_n) converges or the set of its limit points is connected and dense in itself.

2.100 Prove that if $f: [0, 1] \to [0, 1]$ is a continuous function, then the sequence of iterates $x_n = f(x_{n-1})$ converges if and only if $\lim_{n \to \infty} (x_n - x_{n-1}) = 0$.

2.101 Prove that if a sequence $(t_n) \subseteq (0, 1)$ does not converge to zero, then $\prod_{k=1}^{\infty} (1 + (-1)^{x_k} t_k) = 0$ for almost every $x = \sum_{k=1}^{\infty} x_k / 2^k$, $x_k \in \{0, 1\}$.

2.102 Let $s_1 = \log a$, $s_n = \sum_{k=1}^{n-1} \log(a - s_k)$ for $n \geqslant 2$. Show that $\lim_{n \to \infty} s_n = a - 1$.

2.103 Study convergence of the sequence defined by $a_{n+1} = a_n(a_n - 1)$.

2.104 Study convergence of the sequence $a_n = \sum_{k=0}^{[n/2]} \binom{n-k}{k} t^{2n-3k}$.

2.105 Determine when the sequence (a_n) defined by $a_0 = x$, $a_n = x^{2n+1} - \sum_{k=1}^{n} \binom{2n+1}{k} a_{n-k}$ is (i) bounded and (ii) convergent.

2.106 Study for every real x, the convergence of the sequence $u_n = n! \prod_{k=1}^{n} \sin(x/k)$.

2.107 Does the functional sequence $f_n(x) = \int_0^x (1 - t^2)^n \, dt / \int_0^1 (1 - t^2)^n \, dt$ converge?

2.108 Show that no subsequence of the sequence $f_n(x) = nx - [nx]$ converges on a nontrivial interval.

2.109 Calculate $\int_0^1 F_n(x) \, dx$ if $F_0(x) = x$, $F(x) = 4x(1 - x)$, $F_{n+1}(x) = F(F_n(x))$.

2.110 Determine all $x_0 \in [0, 1]$ such that the recurrent sequence $x_{n+1} = 4x_n(1 - x_n)$ converges.

2.111 Compute the integrals $I_n = \int_0^1 f_n(x) \, dx$, where $f_n(x) = (nx - 1)/(1 + x \log n)(1 + nx^2 \log n)$. Explain why $\lim_{n \to \infty} I_n \neq \int_0^1 \lim_{n \to \infty} f_n$.

2.112 Does there exist a sequence of continuous functions $f_n: [0, 1] \to [0, 1]$ such that for every x $f(x) = \lim_{n \to \infty} f_n(x)$ exists but f is discontinuous everywhere?

2.113 Show that the functional sequence $\sin(x + 4\pi^2 n^2)^{1/2}$, $x \geqslant 0$, is precompact in $C(K)$ for every compact $K \subseteq \mathbb{R}^+$ but is not precompact in $C(\mathbb{R}^+)$.

2.114 Construct a continuous bounded function on $[0, 1] \times \mathbb{R}$ which cannot be uniformly approximated by functions of the form $v(x)u(y)$, where v is continuous on $[0, 1]$ and u is continuous and bounded on $\mathbb{R}$.

2.115 Compute $\lim_{n \to \infty} \sup_{x \in \mathbb{R}} ((n + 1)^{1/2} \sin^n x \cos x)$.

2.116 Define the recurrent sequence by $a_1 = x$, $a_{n+1} = \log((\exp(a_n) - 1)/a_n)$. Show that $1 + a_1 + a_1 a_2 + a_1 a_2 a_3 + \cdots = e^x$.

2.117 Suppose that the sequence (a_n) satisfies the conditions: the series $\sum_{n=1}^{\infty} a_n$ converges absolutely and for every natural k $a_k + a_{2k} + a_{3k} + \cdots = 0$. Show that $a_n \equiv 0$.

2.118 Denote by x_k^n the root of the derivative of the polynomial $x(x-1)\cdots(x-n)$ lying in $(k, k+1)$. Determine $\lim_{n \to \infty} x_k^n$ and $\lim_{n \to \infty} (x_n^{2n} - n)$.

2.119 Study the behavior of the recurrent sequence $a_0 = 1$, $a_1 = x$, $a_{n+2} = a_n/(1 + a_{n+1})$.

2.120 Given a complex number z, $|z| \neq 1$, compute $\lim_{n \to \infty} n^{-1} \sum_{k=1}^{n} \sin e^{2\pi i k/n}/(1 - ze^{-2\pi i k/n})$.

2.121 (a_n) is a complex sequence. Prove that the series $\sum_{n=1}^{\infty} a_n b_n$ converges for every complex sequence (b_n) satisfying $\lim_{n \to \infty} |b_n|^{1/n} = 0$ if and only if the sequence $(|a_n|^{1/n})$ is bounded.

2.122 Compute $\lim_{n \to \infty} n^{-2} \sum_{k=1}^{n^2} (n - [\sqrt{(k-1)}])/(\sqrt{k} + \sqrt{k-1})$.

2.123 Let $[\![x]\!]$ denote the distance from x to the integers, $q = (\sqrt{5} + 1)/2$. Show that $\sum_{n=0}^{\infty} [\![q^n]\!] = 1$, $\sum_{n=0}^{\infty} [\![q^{2n}]\!] = 2 \sum_{n=0}^{\infty} [\![q^{3n}]\!]$.

3

Series

Topics covered include:
summation, convergence criteria, functional series, infinite products.

3.1 Prove the Goldbach theorem: If $A = \{k^m : k, m = 2, 3, \ldots\}$ is the set of all natural powers, then $\sum_{n \in A} 1/(n-1) = 1$.

3.2 Show that for the recurrent sequence (a_n) defined by $a_1 = 2$, $a_{n+1} = a_n^2 - a_n + 1$
  (i) All a_n are mutually relatively prime.
  (ii) $\sum_{n=1}^{\infty} a_n^{-1} = 1$.

3.3 Let $S_n = \sum_{k=0}^{n} f_k^2$, where f_k is the kth Fibonacci number, $f_0 = f_1 = 1$, $f_{n+1} = f_n + f_{n-1}$. Compute $\sum_{n=0}^{\infty} (-1)^n/S_n$.

3.4 Show that $\sum_{k=1}^{\infty} (k - n[k/n])/k(k+1) = \log n$, $n \in \mathbb{N}$. Compute $\sum_{k=1}^{\infty} S(n,k)/k(k+1)$, where $S(n,k)$ denotes the sum of all digits in the expansion of k in base n.

3.5 Let $A_k \subseteq \mathbb{N}$ denote the subset of numbers which do not contain the digit k in their decimal expansion. Show that the series $\sum_{n \in A_k} 1/n$ converges. Determine all a such that $\sum_{n \in A_k} 1/n^a$ converges.

3.6 Denote by $f(n)$ the number of zeros in the decimal expansion of n. For which values of a does the series $\Sigma_{n=1}^{\infty} a^{f(n)}/n^2$ converge?

3.7 Prove that $\Sigma_{n=1}^{\infty} (-1)^n \log n/n = (\gamma - (\log 2)/2) \log 2$, where γ is the Euler constant.

3.8 Find the sum of the series $\Sigma_{n=1}^{\infty} (-1)^{n+1} \log(1 + 1/n)$.

3.9 Show that for every sequence (a_n) converging to zero there exists a decreasing sequence (b_n) such that the series $\Sigma a_n b_n$ converges and Σb_n diverges. Similarly show that for every sequence (a_n) diverging to $+\infty$ there exists a decreasing sequence (b_n) such that the series Σb_n converges and $\Sigma a_n b_n$ diverges.

3.10 Let $\Sigma = \{a = (a_n): \text{the series } \Sigma_{n=1}^{\infty} a_n \text{ converges}\}$. Construct the sequence $b \notin \Sigma$ such that $\inf_{a \in \Sigma} \sup_n |1 - a_n/b_n| = 0$ and a sequence $a \in \Sigma$ such that $\inf_{b \notin \Sigma} \sup_n |1 - a_n/b_n| = 0$.

3.11 The terms of two divergent series Σa_n, Σb_n decrease to zero. What can be said about the convergence of the series $\Sigma \min(a_n, b_n)$?

3.12 Let Σa_n be a convergent positive series. Give necessary and sufficient conditions for the existence of a positive sequence (b_n) such that $\Sigma a_n/b_n < \infty$ and $\Sigma b_n < \infty$.

3.13 Prove the Schlömilch theorem: If (g_k) is an increasing sequence of positive integers such that for some $c > 0$ and for all k, $g_{k+1} - g_k \leqslant c(g_k - g_{k-1})$ and the positive sequence (a_n) decreases, then $\Sigma_{n=1}^{\infty} a_n < \infty$ if and only if $\Sigma_{n=1}^{\infty} (g_{n+1} - g_n)a_{g_n} < \infty$.

3.14 Prove that the series $\Sigma_{n=1}^{\infty} a_n$ converges if and only if for every positive sequence (p_n) converging to infinity the sequence $p_n^{-1} \Sigma_{k=1}^{n} p_k a_k$ tends to zero.

3.15 Prove that the series $\Sigma_{n=1}^{\infty} a_n$ is absolutely converging if and only if for every $\varepsilon > 0$ and every increasing sequence of positive integers (m_k) there exists a natural number N such that for all $n > N$, $p \in \mathbb{N}$ $|\Sigma_{k=1}^{p} a_{n+m_k}| < \varepsilon$.

3.16 Let (a_n) be a positive sequence converging to zero and $N(x)$ denote the number of terms in (a_n) greater than x. Show that if $\Sigma_{n=1}^{\infty} a_n < \infty$, then $\lim_{x \to 0} xN(x) = 0$. Is the converse true? Show that $\Sigma_{n=1}^{\infty} a_n = \int_0^{\infty} N(x)\,dx$.

3.17 Let a positive sequence (u_n) increase and $\Sigma_{n=1}^{\infty} 1/u_n < \infty$. Denote by $f(x)$ the number of pairs (i, j) such that $\Sigma_{k=i}^{j} u_k \leqslant x$. Show that $\lim_{x \to \infty} f(x)/x = 0$.

3.18 Determine the functions $f:[0, \infty) \to (0, \infty)$ such that there exists a positive sequence (a_n) converging to zero and satisfying $\Sigma_{n=1}^{\infty} a_n = \infty$, $\Sigma_{n=1}^{\infty} f(a_n) < \infty$.

3.19 Suppose that the positive series $\Sigma_{n=1}^{\infty} a_n$ converges and denote $r_n = \Sigma_{m=n}^{\infty} a_m$. Prove that for every $p \in (0,1)$ there exists a constant C_p satisfying $\Sigma_{n=1}^{\infty} a_n/r_n^p < C_p(\Sigma_{n=1}^{\infty} a_n)^{1-p}$. Find the optimal constant C_p.

3.20 Prove the following generalization of Dini theorem: For a function $F : \mathbb{R}^+ \to \mathbb{R}^+$ the following conditions are equivalent:
  (i) For every positive convergent series $\Sigma_{n=1}^{\infty} a_n$ the series $\Sigma_{n=1}^{\infty} a_n F(r_n)$ converges, $r_n = \Sigma_{m=n}^{\infty} a_m$.
  (ii) There exists a $\delta > 0$ and a decreasing integrable function h on $(0, \delta)$ such that $F \leqslant h$.

3.21 Prove that for a function $F : \mathbb{R}^+ \to \mathbb{R}^+$ the following conditions are equivalent:
  (i) For every positive divergent series $\Sigma_{n=1}^{\infty} a_n$ the series $\Sigma_{n=1}^{\infty} a_n F(s_n)$ converges, $s_n = \Sigma_{m=1}^{n} a_m$.
  (ii) There exists a $\delta > 0$ and a decreasing integrable function h on (δ, ∞) such that $F \leqslant h$.

3.22 Let (λ_n) be a nondecreasing positive sequence and f a nondecreasing positive function such that $\int_{\lambda_1}^{\infty} (tf(t))^{-1} dt < \infty$. Show that $\Sigma_{n=1}^{\infty} (1 - \lambda_n/\lambda_{n+1})/f(\lambda_n) < \infty$.

3.23 (λ_n) is a strictly increasing to infinity sequence and for some $M > 0$ and all $n \in \mathbb{N}$ $\lambda_{n+1} - \lambda_n \geqslant M$ ($\leqslant M$ respectively). A positive decreasing function $f \in C[\lambda_1, \infty)$ satisfies $\int_{\lambda_1}^{\infty} f(t) dt < \infty$ ($= \infty$ respectively). Show that $\Sigma_{n=1}^{\infty} f(\lambda_n) < \infty$ ($= \infty$ respectively).

3.24 Suppose that $f : (0, \infty) \to \mathbb{R}$ is continuous positive and strictly increasing. Show that the series $\Sigma_{n=1}^{\infty} 1/f(n)$ converges if and only if $\Sigma_{n=1}^{\infty} f^{-1}(n)/n^2$ does.

3.25 Prove that the series $\Sigma_{n=2}^{\infty} (\log n)^{-\log \log n}$ diverges.

3.26 Suppose that a real sequence (p_n) decreases to zero, $\Sigma_{n=1}^{\infty} p_n = \infty$ and the signs (ε_n), $\varepsilon_n \in \{-1, 1\}$, are chosen in such a way that $\Sigma_{n=1}^{\infty} \varepsilon_n p_n$ converges. Show that $\lim_{n \to \infty} (\varepsilon_1 + \cdots + \varepsilon_n)p_n = 0$ and $\liminf_{n \to \infty} (\varepsilon_1 + \cdots + \varepsilon_n)/n \leqslant 0 \leqslant \limsup_{n \to \infty} (\varepsilon_1 + \cdots + \varepsilon_n)/n$.

3.27 A complex sequence (z_n) converges to zero. Prove that there exists a choice of signs (ε_n), $\varepsilon_n \in \{-1, 1\}$ such that the series $\Sigma_{n=1}^{\infty} \varepsilon_n z_n$ converges.

3.28 Let σ be a permutation of $\mathbb{N}$ such that $(\sigma(n) - n)$ is a bounded sequence. Show that if the series $\Sigma_{n=1}^{\infty} a_n$ converges, then $\Sigma_{n=1}^{\infty} a_{\sigma(n)}$ does. What can be said in the case of the unbounded sequence $(\sigma(n) - n)$?

3.29 Let σ be a permutation of $\mathbb{N}$ and (x_n) be a real sequence such that:
  (i) $(|x_n|)$ strictly decreases.
  (ii) $\lim_{n \to \infty} |\sigma(n) - n| |x_n| = 0$.

(iii) $\sum_{n=1}^{\infty} x_n = 1$.

Does it imply the equality $\sum_{n=1}^{\infty} x_{\sigma(n)} = 1$? What can be said if (ii) is replaced by the condition

(ii') $\lim_{n \to \infty} |\sigma(n) - n| \, |x_{\sigma(n)}| = 0$?

3.30 Does there exist a positive sequence (b_n) such that $\sum_{n=1}^{\infty} b_{kn} = 1/k$ for all $k = 1, 2, 3, \ldots$?

3.31 Prove that the set of all sums of subseries of a convergent positive series is perfect. Give examples when this set is an interval, a finite union of intervals, and a Cantor set. Describe this set for the series with general terms $a_{2n-1} = 3/4^n$, $a_{2n} = 2/4^n$.

3.32 Let f_0 be a bounded integrable function on $[a, b]$ and $f_n(x) = \int_a^x f_{n-1}(t)\, dt$. Calculate $\sum_{n=1}^{\infty} f_n(x)$.

3.33 Find the limit $\lim_{n \to \infty} \left(\sum_{k=2}^{\infty} k^{-n-1} / \sum_{k=2}^{\infty} k^{-n} \right)$.

3.34 Study convergence of the sequence $s_n(x) = \sum_{k=0}^{n} 2^k \tan(x/2^k) \tan^2(x/2^{k+1})$.

3.35 When does the series $\sum_{n=1}^{\infty} (a^{1/n} - (b^{1/n} + c^{1/n})/2)$, $a, b, c > 0$, converge?

3.36 Study convergence of the series $\sum_{n=1}^{\infty} (-1)^n n^a x^{n^b}$, $a, b, x \in \mathbb{R}$.

3.37 For real a and b, study convergence of the series $\sum_{n=1}^{\infty} (-1)^n / (n^a + 2(-1)^n n^b)$.

3.38 Does the series $\sum_{n=2}^{\infty} (-1)^n \sqrt{n} ((-1)^n + \sqrt{n})^{-1} \sin 1/\sqrt{n}$ converge?

3.39 Study convergence of the series $\sum_{n=1}^{\infty} \sin^n n$.

3.40 Define the sequence (u_n) recurrently by $u_0 = 1$, $u_{n+1} = n^{-p} \sin u_n$, $p \in \mathbb{R}$. Study convergence of the series $\sum_{n=1}^{\infty} u_n$.

3.41 Study convergence of the series $\sum_{n=1}^{\infty} \sin(n! \, x)$, for real x.

3.42 Is the function $f(x) = \sum_{n=1}^{\infty} n^{-1} \sin(x/n)$ bounded for real x?

3.43 Determine the asymptotics of the sum $\sum_{n \leq x} n^{-1}(\sin(x/n) + \sin(n/x))$ when x tends to infinity.

3.44 Show that the series $\sum_{n=1}^{\infty} (-1)^n / (2\sqrt{n} + \cos x)$ converges uniformly on $\mathbb{R}$.

3.45 Show that the function $\sum_{n=1}^{\infty} x \sin nx / (2\sqrt{n} + \cos x)$ is uniformly continuous on compact subsets of $(-2\pi, 2\pi)$.

3.46 Determine all $(x, y) \in \mathbb{R}^2$ such that $\sum_{n=1}^{\infty} x^n \cos ny / \sqrt{n}$ converges.

3.47 Study convergence of the series $\Sigma_{n=1}^{\infty}(-1)^n(\log x)^{1/n}/n$, $x > 1$.

3.48 Let (a_n) be a positive increasing sequence. When does the series $\Sigma_{n=1}^{\infty}(a_n - a_{n-1})/a_n$ converge?

3.49 Find a sequence of natural numbers (n_k) such that $\Sigma_{k=1}^{\infty} 1/n_k < \infty$ but $\Sigma_{k=1}^{\infty} 1/n_{n_k} = \infty$.

3.50 Let K, L be subsets of natural numbers such that $\Sigma_{n\in L} 1/n < \infty$ and $\Sigma_{n\in K} 1/n < \infty$. Is it true that $\Sigma_{n\in K+L} 1/n < \infty$?

3.51 Let (k_n) be an increasing sequence of natural numbers and let u_n denote the least common multiple of $k_1, \ldots, k_n$. Show that $\Sigma_{n=1}^{\infty} 1/u_n < \infty$.

3.52 Study convergence of the series $\Sigma_{n=1}^{\infty} x^{1+1/2+\cdots+1/n}$.

3.53 Find the sum of the series $f(x) = 1/a + x/a(a + d) + \cdots + x^n/a(a + d)\cdots(a + nd) + \cdots$.

3.54 Find the radius of convergence of the series $\Sigma_{n=1}^{\infty} c_1\cdots c_n z^n$, where $c_k > 0$ and $\Sigma_{k=1}^{\infty} c_k/k < \infty$.

3.55 Determine the convergence domains and the domains of absolute convergence for the series $\Sigma_{n=1}^{\infty} x(x - 1)\cdots(x - (n - 1))/n!$, $\Sigma_{n=1}^{\infty} x(x - 1)\cdots(x - (n - 1))/n!n^p$, $\Sigma_{n=1}^{\infty} (ex)^n y(y - 1)\cdots(y - (n - 1))/n^n$.

3.56 Show that $\lim_{x\to 0^+}\Sigma_{n=1}^{\infty}(-1)^{n-1}/n^x = 1/2$.

3.57 Find the limit $\lim_{x\to-\infty}\Sigma_{n=1}^{\infty} x^n/n^n$.

3.58 Show that the function $S(x) = \Sigma_{n=0}^{\infty}(-1)^n/\sqrt{1 + (n + 1)^2 x}$ decreases for $x \in (0, \infty)$ and $\lim_{x\to 0} S(x) = \frac{1}{2}$.

3.59 Determine the value of the product $\Pi_{n=0}^{\infty}(1 + 2^{-2^n})$.

3.60 Find the sum of the series $\Sigma_{n=0}^{\infty} z^{2^n}/\Pi_{k=0}^n (1 + z^{2^k})$.

3.61 Let $a_1 = 1$, $a_n = n(a_{n-1} + 1)$ for $n > 1$. Compute $\Pi_{n=1}^{\infty}(1 + 1/a_n)$.

3.62 Prove that $\frac{e}{2} = (\frac{2}{1})^{1/2}(\frac{2}{3}\cdot\frac{4}{3})^{1/4}(\frac{4}{5}\cdot\frac{6}{5}\cdot\frac{6}{7}\cdot\frac{8}{7})^{1/8}\cdots$.

3.63 (a_n) is a sequence of reals from $(0, 1)$. Show that the series $\Sigma_{n=1}^{\infty} a_n$ diverges if and only if $a_1 + a_2(1 - a_1) + a_3(1 - a_1)(1 - a_2) + \cdots = 1$.

3.64 Show that if the series $\Sigma_{n=1}^{\infty} a_n$ is convergent, $0 < a_n < 1$, then

$$\prod_{n=1}^{\infty}(1-a_n)^{-1} = 1 + \sum_{k=1}^{\infty}\sum_{S_k}(a_{n_1}/(1-a_{n_1}))\cdots(a_{n_k}/(1-a_{n_k})),$$

where Σ_{S_k} denotes the summation over all $n_1, \ldots, n_k$ such that $0 < n_1 < \cdots < n_k$.

3.65 When does the series $\Sigma_{k=1}^{\infty} \Sigma_{I_k} a_{i_1} \cdots a_{i_k}$ converge if $a_i > 0$ and Σ_{I_k} denotes the summation over all $i_1, \ldots, i_k$ satisfying $i_j \in \{1, \ldots, n\}$, $i_1 \neq i_2$, $i_2 \neq i_3, \ldots, i_{k-1} \neq i_k$.

3.66 Show that $\int_0^1 t^{-1} \log t \log(1 - t)\, dt = \Sigma_{n=1}^{\infty} 1/n^3$.

3.67 Show the identity $\int_0^1 (1 + x^p)^{-1}\, dx = \Sigma_{n=0}^{\infty} (-1)^n/(np + 1)$.

3.68 Prove that $\int_0^1 (x - 1)^{-1} x \log x\, dx = 1 - \Sigma_{n=2}^{\infty} 1/(n^2(n - 1))$.

3.69 Show that $\int_0^{\infty} e^{-x} x^{t-1}(1 - e^{-x})^{-1}\, dx = \Gamma(t) \Sigma_{n=1}^{\infty} n^{-t}$.

3.70 Show that for $a > 0$ $\lim_{n \to \infty} \int_0^n (1 - x/n)^n x^{a-1}\, dx = \int_0^{\infty} e^{-x} x^{a-1}\, dx$.

3.71 Compute $\Sigma_{n=0}^{\infty} (-1)^{[(n+1)/2]}(a_n^2 - 1)^{-1}$, where $a_0 = 2$, $a_{n+1} = a_n + (3 + (-1)^n)/2$.

3.72 Show that $\Sigma_{n=1}^{\infty} (2n - 1)!!/n(2n)!! = \log 4$.

3.73 Show that the series $\Sigma_{n=1}^{\infty} (-1)^n (n^{1/n} - 1)$ and $\Sigma_{n=1}^{\infty} (-1)^n (e - (1 + 1/n)^n)$ converge but not absolutely. Determine all p such that the series $\Sigma_{n=1}^{\infty} (n^{1/n} - 1)^p$, $\Sigma_{n=1}^{\infty} (e - (1 + 1/n)^n)^p$ converge.

3.74 Let $f(x) = \Sigma_{n=0}^{\infty} (n!(n + x))^{-1}$, $-x \notin \mathbb{N}$. Compute $f(10)$. Study properties of the function f.

3.75 Let the reals a and y satisfy the conditions $-e^{-1} < a < e^{-1}$, $y = e^{ay}$. Prove that $\Sigma_{n=0}^{\infty} n^n a^n/n! = 1/(1 - ay)$.

3.76 Let $[\![x]\!]$ denote the distance from x to the integers. Compute $\Sigma_{k \in \mathbb{Z}} 2^{-k} [\![2^k x]\!]^2$.

3.77 Compute $\Sigma_{n=1}^{\infty} (-1)^{[2^n x]}/2^n$, $x \geq 0$.

3.78 Find the sum of the series $\Sigma_{n=1}^{\infty} x^{2\{[2^n x]/2\}}/2^n$, where $\{x\}$ denotes the fractional part of x.

3.79 One can prove divergence of the harmonic series $\Sigma_{n=1}^{\infty} 1/n$ by comparison with the series $\Sigma_{n=1}^{\infty} f(n)$, where $f(n) = 2^{-k}$ if $2^{k-1} \leq n < 2^k$. Study convergence of the series $\Sigma_{n=1}^{\infty} (1/n - f(n))$.

3.80 Study convergence of the sequence $u_n = \Pi_{k=2}^n (2 - e^{1/k})$ and the series $\Sigma_{n=2}^{\infty} u_n$.

3.81 Prove that there exist constants C and D such that for $n \geq 2$ $C \log n \leq \Sigma_{k=0}^{\infty} (1 - (1 - 2^{-k})^n) \leq D \log n$.

3.82 Prove that the series $\Sigma_{n=1}^{\infty} (-1)^{[n\sqrt{2}]}/n$ converges.

3.83 Show that the series $\Sigma_{n=1}^{\infty} (-1)^{[\sqrt{n}]}/n^s$ converges for $s > 1/2$ and diverges for $s < \frac{1}{2}$.

3.84 Study convergence of the series $\sum_{n=1}^{\infty} u_n$, $\sum_{n=1}^{\infty} (-1)^n u_n$, where $u_n = \int_0^{\infty} (1 + t^3)^{-n} dt$. Prove existence of a positive number a such that $\lim_{n \to \infty} n^a u_n \neq 0$.

3.85 Study convergence of the series $\sum_{n=1}^{\infty} (-1)^n n^{-a} \int_1^n (x(x + 1)^{1/2})^{-1} dx$, $a \in \mathbb{R}$.

3.86 Study convergence of the series $\sum_{n=1}^{\infty} (n \int_0^{1/2} \sin^2(n\pi x)/\tan \pi x \, dx)^{-1}$.

3.87 Does the series $\sum_{n=1}^{\infty} \int_1^{\infty} \exp(-x^{n^2}) dx$ converge?

3.88 Study the convergence of the series $\sum_{n=0}^{\infty} \int_0^1 \cos(nt^2) dt$.

3.89 Is it possible that the Cauchy product of two conditionally convergent series converge absolutely?

3.90 Determine convergence and divergence of the Cauchy product of the series $\sum_{n=1}^{\infty} (-1)^n n^{-a}$ and $\sum_{n=1}^{\infty} (-1)^n n^{-b}$, $a, b > 0$.

3.91 Let $S_k = \sum_{j=1}^{k} 1/j$, $k_n = \min\{k : S_k \geq n\}$. Calculate $\lim_{n \to \infty} k_{n+1}/k_n$.

3.92 Let us write in the harmonic series p times sign "$+$" then q times sign "$-$" etc. Show that the new series converges if and only if $p = q$.

3.93 Consider the anharmonic series $\sum_{n=1}^{\infty} (-1)^n/n$. Prove the Pringsheim theorem: Consider the permutation of the terms of the anharmonic series such that (without any change of order in the subsequences with even and with odd indexes) among the first n terms of the new series there are p_n positive terms.

Show that the existence of the limit $\lim_{n \to \infty} p_n/n$ is the necessary condition of convergence of new series to a real number or $\pm\infty$. Moreover, if after each group of p positive terms the group of q negative terms follows, then the sum of this series equals $\log 2 + 2^{-1} \log(p/q)$.

3.94 Let $F(n, a) = \int_0^{\infty} xe^{-2x}(x/(x + a))^n dx$. Show that for $a > b > 0$ the series $\sum_{n=0}^{\infty} F(n, a)/F(n, b)$ converges.

3.95 Determine the sum of the series $\sum_{n=0}^{\infty} \binom{2n}{n}^{-1}(4x)^n$, $-1 < x < 1$.

3.96 Give an example of a complex sequence (a_n) such that the series $\sum_{n=1}^{\infty} |a_n|^c$ diverges for every $c > 0$ but the product $\prod_{n=1}^{\infty} (1 + a_n e^{int})$ converges for almost every $t \in (0, 2\pi)$.

3.97 Let $u_{m,n} = (m^2 - n^2)^{-1}$ if $m \neq n$ and $u_{m,m} = 0$. Compute $\sum_{n=0}^{\infty} u_{n,m}$, $\sum_{n=0}^{\infty} \sum_{m=0}^{\infty} u_{n,m}$, $\sum_{n=0}^{\infty} \sum_{m=0}^{\infty} u_{m,n}$.

3.98 Determine $\lim_{m,n \to \infty} \sum_{i=1}^{m} \sum_{j=1}^{n} (-1)^{i+j}/(i + j)$.

3.99 Compute $\sum_{n=1}^{\infty} \sum_{m=1}^{\infty} (mn(m + n + 2))^{-1}$.

3.100   For which real c does the series $\sum_{m=1}^{\infty} \sum_{n=1}^{\infty} |m + ni|^{-c}$ converge?

3.101   Given the double sequence $(x_{m,n})$ denote $S_{m,n} = \sum_{h \le m} \sum_{k \le n} x_{h,k}$. Give an example of a sequence $(x_{m,n})$ such that
   (i) $x_{m,n} = x_{n,m}$.
   (ii) $x_{m,2n} = -x_{m,2n+1} = x_{m+1,2n}$ for $m \ge 2n + 1$.
   (iii) $x_{2n,2n} = 0$.
   (iv) $\lim_{m,n \to \infty} S_{m,n} = 0$.
   (v) the series $\sum_{n=0}^{\infty} x_{m,n}$, $\sum_{m=0}^{\infty} x_{m,n}$ are not convergent.
Construct a sequence $(x_{m,n})$ such that
   (i) $x_{m,n} = 0$ if $|n - m| > 1$.
   (ii) $\sum_{n=0}^{\infty} x_{m,n} = \sum_{m=0}^{\infty} x_{m,n} = 0$ for all m,n.
   (iii) $\lim_{m,n \to \infty} S_{m,n}$ does not exist.

3.102   The density of the subsequence (a_{n_k}) of (a_n) is defined as $\lim_{k \to \infty} k/n_k$. It is well known that if (n_k) is a subsequence of $\mathbb{N}$ and $\sum_{k=1}^{\infty} 1/n_k < \infty$, then (n_k) is of density zero. Construct an increasing sequence of positive numbers (b_n) such that $b_n/n \to \infty$, $\sum_{n=1}^{\infty} 1/b_n = \infty$ with a subsequence of positive density (b_{n_k}) satisfying $\sum_{k=1}^{\infty} 1/b_{n_k} < \infty$.

3.103   Prove the Fejér theorem: If the series $\sum_{n=0}^{\infty} b_n$ is Cesàro summable and $\sum_{n=0}^{\infty} n|b_n|^2 < \infty$, then $\sum_{n=0}^{\infty} b_n$ converges.

3.104   Prove the Kronecker lemma: If (a_n) is a complex sequence and the series $\sum_{n=1}^{\infty} a_n/n$ converges, then $\lim_{N \to \infty} N^{-1} \sum_{n=1}^{N} a_n = 0$.
   Of course the converse is not true. Nevertheless if $\lim_{N \to \infty} N^{-1} \sum_{n=1}^{N} a_n = 0$, then for every $a > 0$ the series $\sum_{n=1}^{\infty} a_n/n^{1+a}$ converges.

3.105   Show that if (p_n) is a sequence of positive numbers such that $\sum_{n=1}^{\infty} 1/p_n < \infty$, then $\sum_{n=1}^{\infty} n^2 p_n/(\sum_{k=1}^{n} p_k)^2 < \infty$.

3.106   Suppose that (a_n) is an increasing sequence of positive numbers such that $\sum_{n=1}^{\infty} 1/a_n = \infty$. Show that the series $\sum_{n=2}^{\infty} (na_n - (n-1)a_{n-1})^{-1}$ is also divergent.

3.107   Prove that if (a_n) is a sequence satisfying the condition $n^{-1} \sum_{k=1}^{n} a_k \ge \sum_{k=n+1}^{2n} a_k$, then $\sum_{k=1}^{\infty} a_k \le 2ea_1$.

3.108 (a_n) is a positive sequence such that $\sum_{n=1}^{\infty} a_n^2 = \infty$. Show that for every $\varepsilon \in (0, 1/2)$ there exists a positive sequence (b_n), $\sum_{n=1}^{\infty} b_n^2 < \infty$, satisfying for every natural N the condition $\sum_{n=1}^{N} a_n b_n > (\sum_{n=1}^{N} a_n^2)^{1/2 - \varepsilon}$.

3.109   Prove the Hardy-Landau inequality: For every positive sequence (a_n) and $p > 1$ $\sum_{n=1}^{\infty} ((a_1^{1/p} + \cdots + a_n^{1/p})/n)^p \le (p/(p-1))^p \sum_{n=1}^{\infty} a_n$.

3.110   Prove the inequality above for $p = -1$.

3.111 Generalize the result of 3.109 proving that for $t > 0$ $\sum_{n=1}^{\infty} (n/(a_1^{-t} + \cdots + a_n^{-t}))^{1/t} < (1 + t)^{1/t} \sum_{n=1}^{\infty} a_n$.

3.112 Prove the Carleman inequality: For every $a_n > 0$ $\sum_{n=1}^{\infty} (a_1 \cdots a_n)^{1/n} < e \sum_{n=1}^{\infty} a_n$.

3.113 Show that for $a_n > 0$ and every natural k $\sum_{n=1}^{\infty} (a_1 \cdots a_n)^{1/n} \leqslant k^{-1} \sum_{n=1}^{\infty} a_n((n + k)/n)^n$.

3.114 Prove that for every bounded positive sequence (t_n) and $a > 0$ the following inequality holds:

$$t_1^{-a} + \sum_{n=1}^{\infty} t_1 \cdots t_n t_{n+1}^{-a} \geqslant \sum_{n=0}^{\infty} (a/(a + 1))^{n-a}.$$

3.115 Suppose that a positive sequence (a_n) satisfies for every natural n $a_n < a_{n+1} + a_{n^2}$. Show that the series $\sum_{n=1}^{\infty} a_n$ diverges.

3.116 Define the function $f : \mathbb{R}^+ \to \mathbb{R}$ by $f(x) = \sum_{k=0}^{\infty} a_k/x^k$, $a_k \geqslant 0$. Study convergence of the series $\sum_{n=1}^{\infty} f(n)$, $\sum_{n=1}^{\infty} \log f(n)$, $\sum_{n=1}^{\infty} (f(1) - f(n))$.

3.117 Show that if a Dirichlet series $\sum_{n=1}^{\infty} a_n/n^z$ converges at $z = z_0$, then it converges uniformly on compacts in the half-plane $\{z : \operatorname{Re} z > \operatorname{Re} z_0\}$.

3.118 Given Dirichlet series $\sum_{n=1}^{\infty} a_n/n^z$ the abscissa of convergence is defined as $\sigma_0 = \inf\{\operatorname{Re} z : \sum_{n=1}^{\infty} a_n/n^z \text{ converges}\}$ and the abscissa of absolute convergence as $\sigma_1 = \inf\{\operatorname{Re} z : \sum_{n=1}^{\infty} |a_n|/n^z \text{ converges}\}$. Show that $\sigma_0 \leqslant \sigma_1 \leqslant \sigma_0 + 1$.

3.119 Prove the Landau theorem: If the abscissa of convergence σ_0 of the Dirichlet series $\sum_{n=1}^{\infty} a_n/n^z$ is finite and $a_n \geqslant 0$ for all n, then $z_0 = \sigma_0 + 0i$ is a singular point of the function $f(z) = \sum_{n=1}^{\infty} a_n/n^z$.

3.120 Show that if two Dirichlet series $\sum_{n=1}^{\infty} a_n/n^z$, $\sum_{n=1}^{\infty} b_n/n^z$ converge in a common half-plane and their sums coincide in an open set, then $a_n = b_n$ for all n.

3.121 Construct the analytic continuation of the zeta function $\sum_{n=1}^{\infty} 1/n^z$ into the domain $\{\operatorname{Re} z > 0\} \setminus \{1\}$.

3.122 Show that the function $f : \mathbb{R} \to \mathbb{C}$ defined by $f(t) = \sum_{n=1}^{\infty} (n(n + 1))^{-1-it}$ does not vanish.

3.123 Let $P_k = \{0, k, 2k, \ldots\}$. Determine $\lim_{x \to \infty} e^{-x} \sum_{n \in P_k} x^n/n!$.

3.124 Compute $\lim_{n \to \infty} e^{-nx} \sum_{k=n}^{\infty} x^k(a + n)^k/k!$.

3.125 Prove that $\sum_{n=1}^{\infty} 1/x_n^2 = 1/10$, where x_ns are positive roots of the equation $\tan x = x$.

3.126 Given a positive constant c and a bounded continuous function $f : \mathbb{R}^+ \to \mathbb{R}^+$ such that $\int_0^\infty f^2(t)\,dt = \infty$, define the sequence $(x_n(c))$ by $\int_0^{x_n(c)} f(t)\,dt = nc$. Study properties of the function $h(c) = \sum_{n=1}^\infty f(x_n(c))$.

3.127 Show that the Maclaurin series of the function $f(x) = \sum_{n=0}^\infty e^{-n} \cos n^2 x$ converges only at $x = 0$.

3.128 Show that the function $f(x) = \sum_{n=1}^\infty x^n/n^2$ satisfies for $x \in (0, 1)$ the identity $f(x) + f(1 - x) + \log x \cdot \log(1 - x) = f(1)$.

3.129 Study convergence of the infinite product $\prod_{n=1}^\infty (1 + x^n/n^p) \cos(x^n/n^q)$.

4

Functions of One Real Variable

Topics covered include:
Inequalities, integrals, continuous and differentiable functions

4.1 Show that for arbitrary real numbers $a, b > 0$ and natural number n
$\Pi_{k=1}^{n} (a^k + b^k)^2 \geq (a^{n+1} + b^{n+1})^n$.

4.2 Prove the inequality

$$\frac{\prod\limits_{j=1}^{n} x_j}{\left(\sum\limits_{j=1}^{n} x_j \right)^n} \leq \frac{\prod\limits_{j=1}^{n} (1 - x_j)}{\left(\sum\limits_{j=1}^{n} (1 - x_j) \right)^n}$$

for all $x_j \in (0, 1/2]$, $j = 1, \ldots, n$, and natural n.

4.3 Show that for all $r_{ij} \in [0, 1]$, $i = 1, \ldots, m, j = 1, \ldots, n$, the inequality

$$1 - \prod_{j=1}^{n} \left(1 - \prod_{i=1}^{m} r_{ij} \right) \leq \prod_{i=1}^{m} \left(1 - \prod_{j=1}^{n} (1 - r_{ij}) \right)$$

holds.

4.4 Prove for positive numbers $x_j, j = 1, \ldots, n$, the inequalities

$$\left(\prod_{j=1}^{n} x_j\right)^{\Sigma_{j=1}^n x_j/n} \leqslant \prod_{j=1}^{n} x_j^{x_j} \leqslant \left(\sum_{j=1}^{n} x_j^2 \bigg/ \sum_{j=1}^{n} x_j\right)^{\Sigma_{j=1}^n x_j}$$

4.5 Define for positive numbers $x_1, \ldots, x_n$ the function $R(x_1, \ldots, x_n) = (x_1 + \cdots + x_n)^{x_1 + \cdots + x_n}/x_1^{x_1} \cdots x_n^{x_n}$. Show that $R(x_1, \ldots, x_n)R(y_1, \ldots, y_n) \leqslant R(x_1 + y_1, \ldots, x_n + y_n)$.

4.6 Prove that if $H(t) = t^t$, $x = x_1 + \cdots + x_n$ and $x_j > 0$, then the inequalities

$$H(1 + x)\bigg/\prod_{i=1}^{n} H(1 + x_i) \leqslant \Gamma(1 + x)\bigg/\prod_{i=1}^{n} \Gamma(1 + x_i) \leqslant H(x)\bigg/\prod_{i=1}^{n} H(x_i)$$

hold.

4.7 Show that for $0 < x_j < \pi$ and $x = (x_1 + \cdots + x_n)/n$,

$$\prod_{j=1}^{n} \sin x_j/x_j \leqslant (\sin x/x)^n.$$

4.8 Define for x, $y > 0$ and r, $s > 0$ the means $M_1(x, y) = ((x^r y^s + x^s y^r)/2)^{1/(r+s)}$, $M_2(x, y) = ((x^{rs} + y^{rs})/2)^{1/rs}$. Determine the parameters r and s such that $M_1(x, y) < M_2(x, y)$ holds for all $x \neq y$ and determine when the means M_1 and M_2 are not comparable.

4.9 Define the logarithmic mean of positive numbers x and y by $L(x, y) = ((\log x - \log y)/(x - y))^{-1}$ if $x \neq y$ and $L(x, x) = x$. Recall the definition of the power means $m_p(x, y) = ((x^p + y^p)/2)^{1/p}$ if $p \neq 0$ and $m_0(x, y) = \sqrt{xy}$. Show that $m_p(x, y) \leqslant L(x, y) \leqslant m_q(x, y)$ holds for all positive x, y if and only if $p \leqslant 0$ and $q \geqslant 1/3$.

4.10 Determine all positive functions f and g on $\mathbb{R}^2$ such that $(\Sigma_{j=1}^n a_j b_j)^2 \leqslant \Sigma_{j=1}^n f(a_j, b_j) \Sigma_{j=1}^n g(a_j, b_j) \leqslant (\Sigma_{j=1}^n a_j^2)(\Sigma_{j=1}^n b_j^2)$ holds for all positive a_j, b_j.

4.11 Let $e_0 = 0$, $e_k = \exp e_{k-1}$. Show that for arbitrary reals $t_1, \ldots, t_n$ and natural n, the inequality $\Sigma_{j=1}^n (1 - t_j) \exp(\Sigma_{k=1}^j t_k) \leqslant e_n$ is true with the given equality if and only if $t_n = e_0$, $t_{n-1} = e_1, \ldots, t_1 = e_{n-1}$.

4.12 Prove that if $n \geqslant 2$ and $x_j > 0$ for $j = 1, \ldots, n$, then $x_1^{x_2} + x_2^{x_3} + \cdots + x_{n-1}^{x_n} + x_n^{x_1} \geqslant 1$.

4.13 Show that for positive numbers $x_j, j = 1, \ldots, n, n \geqslant 2$ and $r \leqslant n$

$$r \sum_{1 \leqslant j_1 < \cdots < j_r \leqslant n} \frac{x_{j_1} \cdots x_{j_r}}{x_{j_1} + \cdots + x_{j_r}} \leqslant \binom{n}{r}\left(\frac{x_1 + \cdots + x_n}{n}\right)^{r-1}$$

4.14 Assume that $x_1, \ldots, x_n$, $n \geqslant 2$, are in $[0, 1]$. Find the best upper estimate of $S_n = (x_1 - x_2)^2 + (x_2 - x_3)^2 + \cdots + (x_n - x_1)^2$ and determine the conditions when the equality in your estimate holds.

4.15 Show that for mutually distinct reals $x_1, \ldots, x_n$ $\sum_{k=1}^n ((1 + x_k^2)^{n/2} / \prod_{j \neq k} |x_j - x_k|) \geqslant n$. When does the equality hold?

4.16 Show that $\sum_{k=2}^n a_k^{1-1/k} < s + 2\sqrt{s}$, for arbitrary positive $a_2, a_3, \ldots, a_n$ and $s = a_2 + a_3 + \cdots + a_n$.

4.17 Let $s = x_1 + \cdots + x_n$ for positive x_j, $n \geqslant 2$ and $0 < c \leqslant 1$. Prove that $\sum_{k=1}^n ((s - x_k)/x_k)^c \geqslant (n - 1)^{2c} \sum_{k=1}^n (x_k/(s - x_k))^c$ with the equality valid if and only if $x_1 = \cdots = x_n$.

4.18 Show that for $n \geqslant 2$ and $0 < x < n/(n + 1)$ $(1 - 2x^n + x^{n+1})^n < (1 - x^n)^{n+1}$.

4.19 Show the inequality $\sum_{m=0}^{[x]} (x - m)^p \leqslant (x + 1/2)^{p+1}/(p + 1)$ valid for all $x \geqslant 0$ and $p \geqslant 1$. When does the equality hold?

4.20 Prove that $\inf_{0 \leqslant x \leqslant 1} \sum_{j=1}^n |x - p_j|^{-1} \leqslant 8n(1 + \frac{1}{3} + \cdots + 1/(2n - 1))$ for arbitrary $0 \leqslant p_j \leqslant 1$.

4.21 Show that for all natural m and n, and real $a > 1$ $\sum_{k=1}^{n^m - 1} a^k [k^{1/m}] \leqslant (n - 1)(a^{n^m} - a^{n^m/2^m})/(a - 1)$.

4.22 Assume that for some $a_k > -1$, $k = 1, \ldots, n$, and every $0 \leqslant x \leqslant 1$ the inequality $a_1/(1 + a_1 x) + \cdots + a_n/(1 + a_n x) \leqslant 1$ is verified. Show that $(1 + a_1) \cdots (1 + a_n) \leqslant e$.

4.23 Determine $\lim_{x \to \infty} ((a^x - 1)/(x(a - 1)))^{1/x}$ for $a > 0$, $a \neq 1$.

4.24 Compute $\lim_{x \to \infty} x(1/e - (x/(x + 1))^x)$.

4.25 Does the limit of $\sqrt{1 - x} \prod_{n=0}^\infty (1 + x^{4^n})$ exist as $x \to 1 - 0$?

4.26 Compute $\lim_{x \to 1-0} (1 - x)^{p+1} \sum_{n=1}^\infty n^p x^n$.

4.27 Let $S_n = 1 + \frac{1}{2} + \cdots + 1/n$, $f(x) = \sum_{n=1}^\infty a_n x^n$. Give an explicit formula for the function $g(x) = \sum_{n=1}^\infty a_n S_n x^n$ using the function f.

4.28 Determine all positive real a such that $a^x \geqslant ax$ holds for all positive x.

4.29 Prove the inequalities $n(n - 1)^{-1} \log n < \sum_{k=1}^n 1/k < \log n + n \log(1 + 1/n)$.

4.30 Show that for $a > 1$ and $x > 0$ $-\log(1 - (1 - e^{-x})^a) < x^a$.

4.31 Define for $x > 1$ $g(x) = \lim_{n \to \infty} (\log(x + n) - \sum_{k=0}^n (x + k)^{-1})$, $h(x) = \lim_{n \to \infty} (\log \log(x + n) - \sum_{k=0}^n ((x + k) \log(x + k))^{-1})$. Prove that for sufficiently large x $h(x) > \log g(x)$.

4.32 Prove that for all real x $\lim_{t \to \infty} \prod_{n=0}^{\infty} [1 + tx^2/(n + t)^2] = e^{x^2}$.

4.33 Show that the expression $\exp(-x^2/2) \sum_{k=0}^{\infty} x^{2k+1}/(2k + 1)!!$ converges monotonically to $\sqrt{\pi/2}$ as x tends to infinity.

4.34 Prove that $2 \arctan (1/(2n - 1)) < \sum_{k=n}^{\infty} 1/k^2$.

4.35 Show the inequality $|\pi/4 - \arctan x| \leqslant \dfrac{\pi}{4} \dfrac{|x - 1|}{(x^2 + 1)^{1/2}}$ for arbitrary positive x.

4.36 Show that for $0 \leqslant t \leqslant \pi/2$, $\sin t \geqslant \dfrac{2}{\pi} t + \dfrac{t}{12\pi} (\pi^2 - 4t^2)$.

4.37 Show that for $0 < x < \sqrt{\pi/2}$, $(\sin x)^2 \leqslant \sin x^2$.

4.38 Is the inequality $a \tanh x > \sin ax$ true for all $x > 0$ and $a > 1$?

4.39 Prove that $\cos x + \cos y \leqslant 1 + \cos xy$ for all x and y satisfying $0 \leqslant x^2 + y^2 \leqslant \pi$.

4.40 Show that for $0 < v < \pi/2$ and $0 < u < \pi$, $\cos(u \sin v) \cos(u \cos v) - \cos u > 0$. Determine $\lim_{v \to 0+} u(v)$, where $u(v)$ denotes the least positive u such that the expression above vanishes.

4.41 Solve the equation $\sqrt{\sin x} + \sqrt{\cos x} = t$.

4.42 Prove that for every real a the equation $x = -a + \sqrt{2} \sin((a - x)/\sqrt{2})$ has exactly one solution.

4.43 Determine the number of solutions of the equation $a^x = \log_a x$, where $a > 0$, $a \neq 1$.

4.44 Prove that for $k \geqslant 3$ the equation $(\log x)^k = x$ has exactly two solutions r_k, s_k in $[1, \infty)$ such that $r_k \to e$, $s_k \to \infty$ as $k \to \infty$.

4.45 Suppose that for a function $f: \mathbb{R} \to \mathbb{R}$, $g(a) = \lim_{t \to \infty} f(at)t^{-c}$ is finite for all $a > 1$ and some real c. Show that $g(a) = Da^c$ for some constant D.

4.46 Show that $\dfrac{1}{2} \displaystyle\int_1^2 (-1)^t e^{(-1)^t x} dt = \dfrac{\sinh x}{i\pi x}$, where $(-1)^t$ denotes $e^{i\pi t}$.

Observe an analogy with the formula $\frac{1}{2} \sum_{t=1}^2 (-1)^t e^{(-1)^t x} = \sinh x$.

4.47 Compute $\lim_{t \to 0+} (\int_0^1 (bx + a(1 - x))^t dx)^{1/t}$ for $0 < a < b$.

4.48 Let $[\![x]\!]$ denote the distance from x to the set of integers. Compute $\lim_{n \to \infty} \frac{1}{n} \int_1^n [\![n/x]\!] dx$.

4.49 Study continuity and differentiability of the function $f(x) = \int_0^{\infty} e^{-xt^2}/(1 + t^2) dt$.

4.50 Study convergence of the integral $\int_0^\infty x \sin(x^3 - ax)\, dx$ for $a > 0$.

4.51 Compute $\displaystyle\int_0^\infty \frac{\arctan(\pi x) - \arctan x}{x}\, dx$.

4.52 Show that $\lim_{n \to \infty} \int_0^1 x^{-1}(1 - (1 - x/n)^n)\, dx = \int_0^1 x^{-1}(1 - e^{-x})\, dx$.

4.53 Determine the limit $\lim_{t \to 0} \sqrt{t} \int_1^\infty dx/(\sqrt{x}(1 + tx))$.

4.54 Compute $\lim_{n \to \infty} \int_{2n\pi}^{(2n+1)\pi} (x^a - [x^a]) \sin x\, dx$ if $a > 1$.

4.55 Show that for $p > 0$ and $x \geqslant 0$ $\frac{2}{\pi} \int_x^{px} (\sin t/t)^2\, dt \leqslant 1 - 1/p$.

4.56 Compute the integral $\int_{-1}^1 dx/((e^x + 1)(x^2 + 1))$.

4.57 Determine the set $W = \{ \int_0^1 f(x)e^{i\pi x}\, dx : f \in C[0,1], |f(x)| \leqslant 1 \}$.

4.58 Study the asymptotics of the function $f(t) = \int_0^t x^t \sin x\, dx$ as t tends to infinity.

4.59 Find the asymptotics of the integral $\int_0^\infty \sin t^n\, dt$, $n \to \infty$.

4.60 Study continuity and differentiability of the function $I(x) = \int_0^\infty (e^t - 1)^{-1} \sin xt\, dt$. Show that $I(x) = \sum_{n=0}^\infty x/(x^2 + n^2)$.

4.61 Show that if $f(t) = \prod_{n=1}^\infty (1 + t/n^a)$, $a > 1$, then $\lim_{t \to \infty} t^{-1/a} \log f(t) = \pi(\sin \pi/a)^{-1}$.

4.62 Compute the product $\prod_{k=1}^{n-1} \sin(k\pi/2n)$ and then the integral $\int_0^{\pi/2} \log(\sin x)\, dx$.

4.63 Find $\int_0^\infty (\pi^2 - x^2)^{-1} \sin^2 x\, dx$.

4.64 Compute $\lim_{n \to \infty} \int_0^\pi (1 + \cos^2 nx)^{-1} \sin x\, dx$.

4.65 Let $R(a,p) = \int_0^\infty e^{-ax}(1 + x)^p\, dx$. Determine the conditions on $a, b > 0$, $p, q > 0$, guaranteeing that $1/R(a, p) + 1/R(b, q) \geqslant 1/R(a + b, p + q)$.

4.66 Prove that if the functions g and h satisfy $g(1/x) = g(x)$, $h(1/x) = -h(x)$ for $x > 0$, then $\int_0^\infty g(x)(1 + h(x))x^{-1}\, dx = \int_0^\infty g(x)x^{-1}\, dx$. In particular show that the integral $\int_0^\infty (1 + x^a)^{-1}(1 + x^2)^{-1}\, dx$ does not depend on a.

4.67 Prove the asymptotic formula

$$\text{p.v.} \int_{-x}^\infty \frac{e^{-x-t}}{t}\, dt \simeq -\left(\frac{1}{x} + \frac{1}{x^2} + \frac{2!}{x^3} + \cdots + \frac{(n-1)!}{x^n} + \cdots \right)$$

as $x \to \infty$.

4.68 Show that $(\int_0^a e^{-x^2}\, dx)^2 = \int_0^{\pi/4} (1 - e^{-a^2/\cos^2 t})\, dt$. Can one take the limit as $a \to \infty$ in this formula?

4.69 Suppose that a function $f:\mathbb{R}^+ \to \mathbb{R}^+$ satisfies for all real s the estimate $\int_0^\infty f(x)e^{sx}\, dx \leq \exp(s^2)$. Show that there exists $\varepsilon > 0$ such that $\int_0^\infty f(x) \exp(\varepsilon x^2)\, dx < \infty$.

4.70 Find an upper bound for the quotient $\int_0^{\pi/2-\varepsilon} \sin^n x\, dx / \int_0^{\pi/2} \sin^n x\, dx$.

4.71 Find the function $f(x)$ of the form $(x+b)/(cx+d)$ such that $\int_0^{\pi/2} |\tan x - f(x)|\, dx$ attains its minimal value.

4.72 Prove that the function $f(x) = \displaystyle\int_{-\infty}^\infty \left(\frac{\sin t}{t}\right)^{2k} \cos 2tx\, dt$, $k \in \mathbb{N}$, satisfies

(i) $f(x) = 0$ for $|x| \geq k$ and $f(x) > 0$ elsewhere.
(ii) f is a polynomial of degree at most $2k - 1$ in every interval of the form $(j, j+1), j = -k, -k+1, \ldots, k-1$.

4.73 Study continuity properties of the function defined by the series $\sum_{n=1}^\infty x/(x + n^2 x^2)$.

4.74 Show that if f is uniformly continuous on $[1, \infty)$ then $f(x) = O(x)$.

4.75 Let f be an increasing continuous function on $\mathbb{R}$ such that $f(x) - x$ is periodic of period 1 and let f^n denote the nth iteration of f. Prove that the limit $\lim_{n \to \infty} f^n(x)/n$ exists. Verify that the continuity assumption is essential.

4.76 Let u_n denote the unique positive root of the polynomial $x^n + x^{n-1} + \cdots + x - 1$. Determine $\lim_{n \to \infty} u_n$.

4.77 Let $0 < a_1 < \cdots < a_n$ be the roots of the polynomial $x^n + c_{n-1}x^{n-1} + \cdots + c_1 x + c_0$. Show that for every $m \leq n$ the polynomial $c_m x^m + \cdots + c_0$ is of constant sign on $[a_m, \infty)$.

4.78 Suppose that the coefficients of the polynomial $P(x) = a_n x^n + \cdots + a_1 x + a_0$, $n \geq 2$, satisfy the inequality $0 < a_0 < -\sum_{k=1}^{[n/2]} a_{2k}/(2k+1)$. Show that P has a real root in $(-1, 1)$.

4.79 Prove that the polynomials P_n defined inductively by $P_0 = 1$, $P_1 = x + 1$, $P_{n+1} = P_n + xP_{n-1}$, have only real roots.

4.80 Is it possible that a function $f(x) = \sum_{k=1}^n a_k (\cos b_k x + c_k)$ with $\prod_{k=1}^n a_k b_k \neq 0$ has a constant sign?

4.81 Show that for arbitrary reals $a_0, \ldots, a_n$, $x_0, \ldots, x_n$, the function $f(x) = a_0 + \sum_{j=1}^n a_j \cot an(x - x_j)$ changes sign in $(0, \pi)$ at most $2n$ times. Moreover, show that for fixed $0 < t_1 < \cdots < t_{2n} < \pi$ the numbers a_j, x_j can be chosen in such a way that f changes sign exactly at $t_1, \ldots, t_{2n}$.

4.82 A continuous function f on the interval $[0, n]$, $n \in \mathbb{N}$, satisfies $f(0) = f(n)$. Show that the equation $f(x) = f(y)$ has at least n solutions with integer $x - y$.

4.83 Prove that a continuous function f on $[0, \pi]$ satisfying $\int_0^\pi f(t) \sin t\, dt = \int_0^\pi f(t) \cos t\, dt = 0$ has at least two zeros on $(0, \pi)$.

4.84 Prove that if f is a continuous function on the half-line $[0, \infty)$ and $\lim_{n \to \infty} f(nh) = 0$ for all h from an open interval, then $\lim_{x \to \infty} f(x) = 0$.

4.85 Suppose that $f : \mathbb{R} \setminus \{0\} \to \mathbb{R}$ satisfies for every a the condition $\lim_{n \to \infty} f(a/n) = 0$. Does it imply that $\lim_{x \to 0} f(x) = 0$?

4.86 Show that if $f : \mathbb{R} \to \mathbb{R}$ satisfies the conditions $f(0) > 0$, $f(1) < 0$ and there exists a continuous function g such that $f + g$ is nondecreasing, then $f(x) = 0$ for some x.

4.87 A continuous function $f : [0, 1] \to \mathbb{R}$ attains each of its values finitely many times and $f(0) \neq f(1)$. Show that f admits some of its values an odd number of times.

4.88 Show that if a homeomorphism $f : \mathbb{R} \to \mathbb{R}$ satisfies the condition $f^n(x) = x$ for some natural n and all x, then
   (i) $f(x) = x$ if f is orientation preserving.
   (ii) $f^2(x) = x$ if f changes orientation.

4.89 Find all functions $f : \mathbb{R} \to \mathbb{R}$ with Darboux property such that $f^m(x) = -x$ for some $m \geqslant 1$ and all x.

4.90 Consider a continuous decreasing function f on the real line. Show that
   (i) f has a unique fixed point.
   (ii) The set $\{x : f^2(x) = x\}$ is infinite or it has an odd number of elements. For fixed $n > 2$ study the structure of the set $\{x : f^n(x) = x\}$.

4.91 f is a continuous mapping of a compact $K \subseteq \mathbb{R}$ into itself and $x_0 \in K$ has the following property: each limit point of the sequence of iterates $(f^n(x_0))$ is a fixed point of f. Show that the sequence $(f^n(x_0))$ is convergent. Give an example that this is not true in the two-dimensional case.

4.92 f is a continuous mapping of a topological space. Show that if the nth iteration of f has a unique fixed point x_0, then $f(x_0) = x_0$.

4.93 Show that for every function $f : [a, b] \to [a, b]$ satisfying $|f(x) - f(y)| \leqslant |x - y|$ the sequence (x_n) defined by $x_{n+1} = (x_n + f(x_n))/2$ converges to a fixed point of f.

4.94 Define for a continuous function $f : [0, 1] \to [0, 1]$ $C(f) = \{x :$ the sequence $(f^n(x))$ is eventually periodic$\}$ and $L_x =$ the set of limit points of the orbit of x. Prove that for every $x \in [0, 1]$ $L_x \in \overline{C(f)}$. Does it remain true in the two-dimensional case?

4.95 Suppose that f and g are continuous functions, one of them being monotone in (a, b) and there exists a sequence (x_n) such that $f(x_n) = g(x_{n-1})$. Show that the equation $f(x) = g(x)$ has a solution in (a, b).

4.96 A real function f is defined on a separable metric space. Show that the set of a, such that $\lim_{x \to a} f(x)$ exists and it is different from $f(a)$, is at most countable.

4.97 Prove that if $f: \mathbb{R} \to \mathbb{R}$ has the Darboux property and $f^{-1}(\{q\})$ is closed for every rational q, then f is continuous.

4.98 Give examples of a function $f: \mathbb{R} \to \mathbb{R}$ with its graph:
  (i) dense in $\mathbb{R}^2$ and connected
  (ii) connected, but f is discontinuous at every rational point and continuous at the irrational points

4.99 Consider a nonnegative function f, continuous at the origin, concave on the interval $(0, 1)$ and periodic of period 1. Prove that $f(nx) \leqslant nf(x)$ for all real x and natural n.

4.100 Suppose that a function $f: \mathbb{N} \to \mathbb{R}$ satisfies the subadditivity condition: $f(a + b) \leqslant f(a) + f(b)$ for all $|b - a| \leqslant k$, where k is a fixed number. Does there exist the limit $\lim_{n \to \infty} f(n)/n$?

4.101 Show that if a continuous function $f: \mathbb{R}^+ \to \mathbb{R}$ satisfies $f(x) \leqslant f(nx)$ for all $x \in \mathbb{R}$, $n \in \mathbb{N}$, then $\lim_{x \to \infty} f(x)$ exists (and may be infinite).

4.102 A function $f: \mathbb{R} \to \mathbb{R}$ continuous from the right satisfies the condition $\lim_{n \to \infty} (\max_{k \in \mathbb{Z}} |f((k + 1)/n) - f(k/n)|) = 0$. Must it be continuous?

4.103 Given a continuous function f on $[0, 1]$ show that $\lim_{n \to \infty} n^{-1} \sum_{k=1}^{n} (-1)^k f(k/n) = 0$.

4.104 Define for a bounded real function f on $[0, 1]$ $Sf(x) = \lim \sup_{\varepsilon \to 0} \{|f(y) - f(z)| : |x - z| < \varepsilon, |x - y| < \varepsilon\}$. Prove that
  (i) Sf is upper semicontinuous.
  (ii) f is continuous at x_0 if and only if $Sf(x_0) = 0$.
  (iii) $S^3 = S^2$.

4.105 Show (without use of the de l'Hôpital rule) that if $\lim_{x \to 0} f(x) = 1$, $\lim_{x \to 0} g(x) = \infty$, $\lim_{x \to 0} g(x) \cdot (f(x) - 1) = c$, then $\lim_{x \to 0} f(x)^{g(x)} = e^c$.

4.106 Define for a positive continuous function f on the interval $[a, b]$ the geometric mean $M_0(f, [a, b]) = \lim_{n \to \infty} (\Pi_{j=0}^{n} f(a + j(b - a)/n))^{1/(n+1)}$. Compute $M_0(x^m, [k, k + 1])$, $m, k = 1, 2, \ldots$, $M_0(c^x, [a, b])$.

4.107 Show that if f is continuous on $[a, b]$, $p_1, \ldots, p_n > 0$, then for every $x_1, \ldots, x_n \in [a, b]$ there exists $x \in [a, b]$ such that $p_1 \int_x^{x_1} f + \cdots + p_n \int_x^{x_n} f = 0$.

4.108 A decreasing function $f:(0,1] \to \mathbb{R}$ has a convergent integral $\int_0^1 x^a f(x)\, dx$ for some a. Show that $\lim_{x \to 0} x^{a+1} f(x) = 0$.

4.109 Let $f: \mathbb{R}^+ \to \mathbb{R}^+$ be a uniformly continuous function. Is it true that $\lim_{x \to \infty} f(x + 1/x)/f(x) = 1$?

4.110 Prove that for a real function defined on an open interval the following properties are equivalent:
   (i) f is uniformly continuous.
   (ii) $\forall \varepsilon > 0 \, \exists N > 0 | (f(x) - f(y))/(x - y)| > N \Rightarrow |f(x) - f(y)| < \varepsilon$.

4.111 A sequence of functions (f_n) is said to be convergent to f quasi-uniformly (or Borel convergent) on a set A if

$$\forall p \in \mathbb{N} \forall \varepsilon > 0 \exists q > p \forall x \in A \exists n(x) \in \mathbb{N}: p < n(x) < q \wedge |f(x) - f_{n(x)}(x)| < \varepsilon.$$

Prove the Arzelà theorem: The pointwise limit of a sequence of continuous functions defined on a compact set is a continuous function if and only if this sequence converges quasi-uniformly.

4.112 Give an example of a sequence of continuous functions on $\mathbb{R}$ converging pointwise to zero which is not uniformly convergent on any nonempty open set.

4.113 Given a function f such that $f(0) = 0$, $f(1) \neq 0$ and $\lim_{n \to \infty} f(n) = 0$, construct a sequence of integers $a_n \to \infty$ and a compact set K such that $f(a_n x) \to 0$ but this convergence is not uniform on K.

4.114 (f_n) is a decreasing sequence of nonnegative continuous functions on $[0,1]$ such that if f is continuous and for every n $f_n \geqslant f \geqslant 0$, then $f \equiv 0$. Does it imply that $\int_0^1 f_n \to 0$?

4.115 Let $P_n(x) = x^n(bx - a)^n/n!$, where $a, b, n \in \mathbb{N}$. The polynomial P_n and its derivatives at $x = 0$ and $x = a/b$ vanish or they are natural numbers. Show that the integrals $\int_0^\pi P_n(x) \sin x\, dx$, $\int_0^r P_n(x)e^x\, dx$ converge to zero as n tends to infinity. Deduce that π and e^r for rational r are irrational.

4.116 Show that any continuous function f on $[0,1]$ such that $f(0) = 0$ can be approximated by polynomials of the form $\Sigma\, w_{k,n} x^k$ satisfying the condition $\lim_{n \to \infty} w_{k,n} = 0$ for every k.

4.117 Prove that the polynomials approximating a continuous function f on $[0,1]$ can be chosen in such a way that their values at $t_1, \ldots, t_n \notin [0,1]$ coincide with given numbers $y_1, \ldots, y_n$.

4.118 Show that a function satisfying the Lipschitz condition can be uniformly approximated on compact sets in $\mathbb{R}$ by polynomials with the same Lipschitz constant.

4.119 Prove that if for continuous functions f and g on $[0,1]$ the equality $f(y_1) = f(y_2)$ implies $g(y_1) = g(y_2)$, then there exists a sequence of polynomials (p_n) such that $p_n(f) \to g$ uniformly on $[0,1]$.

4.120 Construct a complex sequence (a_n) such that for any continuous function $f : [0,1] \to \mathbb{C}$, $f(0) = 0$, there exists a subsequence of indexes (n_k) such that $\sum_{n=1}^{n_k} a_n x^n \to f$ uniformly on $[0,1]$.

4.121 Show that if $\sum_{n=1}^{\infty} a_n = s$ and $g(t) = \sum_{n=1}^{\infty} a_n t^n / n!$, then $\int_0^{\infty} e^{-t} g(t)\, dt = s$.

4.122 Prove that for every $f \in C^1[-a, a]$
$$\lim_{\lambda \to \infty} \int_{-a}^{a} \frac{1 - \cos \lambda x}{x} f(x)\, dx = \int_0^a \frac{f(x) - f(-x)}{x}\, dx.$$

4.123 Let $f(x) = \sum_{n=1}^{\infty} a_n x^n$, $s_m = \sum_{n=1}^{m} a_n x^n$ and $r \neq 0$ is an interior point of the convergence interval of the power series f. Prove that if $s_m(r) < f(r)$ for $m = 1, 2, \ldots$ then $f'(r) \neq 0$.

4.124 (x_n) is a sequence of reals in $[0, 1)$, $A_N(p, k)$ denotes the number of elements of $\{x_1, \ldots, x_N\}$ in the interval $[(k-1)/p, k/p)$, $k = 1, \ldots, p$. Prove that if for every p and $k, l \in \{1, \ldots, p\}$ $\lim_{N \to \infty} A_N(p, k)/A_N(p, l) = 1$, then the sequence (x_n) is uniformly distributed in $[0, 1]$.

4.125 Prove that for $0 \leqslant x_n < 1$ and $P_n(z) = \prod_{k=1}^{n} (z - \exp(2\pi i x_k))$ the following conditions are equivalent:
   (i) (x_n) has uniform distribution on $[0, 1]$.
   (ii) $\lim_{n \to \infty} (\max_{|z|=1} |P_n(z)|)^{1/n} = 1$.
   (iii) $\lim \sup_{n \to \infty} |P_n(e^{it})|^{1/n} = 1$.
   (iv) $\lim_{n \to \infty} |P_n(z)|^{1/n} = 1$ uniformly on compact subsets of $\{z : |z| < 1\}$.
   (v) $\lim_{n \to \infty} |P_n(z)|^{1/n} = |z|$ uniformly on compact subsets of $\{z : |z| > 1\}$.

4.126 Prove the Fejér theorem: If f, g are continuous functions on $\mathbb{R}$ of period 1, then $\lim_{n \to \infty} \int_0^1 f(x) g(nx)\, dx = \int_0^1 f(x)\, dx \cdot \int_0^1 g(x)\, dx$.

4.127 Show that for a positive continuous function g on $\mathbb{R}^+$ the function $f(t) = g(t) \int_0^t g(s)^{-a}\, ds$, $a > 1$, is unbounded. What can be said about f in the case $a = 1$?

4.128 Show that if f is a nonnegative continuous function on $\mathbb{R}^+$, $\int_0^{\infty} f = \infty$, then there exists h such that $\sum_{n=1}^{\infty} f(nh) = \infty$.

4.129 Show that for a positive function f decreasing on $[0, 1]$
$$\frac{\int_0^1 x(f(x))^2\, dx}{\int_0^1 x f(x)\, dx} \leqslant \frac{\int_0^1 (f(x))^2\, dx}{\int_0^1 f(x)\, dx}.$$

4.130 Determine the limit $\lim_{n \to \infty} \left(\int_0^1 x^n f(x)\, dx \right) / \left(\int_0^1 x^n g(x)\, dx \right)$ for the continuous functions f, g on $[0, 1]$, $g > 0$.

4.131 Show that for any positive function f on $[a, b]$ and fixed real u the expression $(\int_a^b f^{s+u}/\int_a^b f^s)^{1/u}$ increases with s.

4.132 The continuous positive functions h and g on $\mathbb{R}^+$ satisfy the conditions $\int_0^\infty g < \infty$ and $\int_0^\infty h = \infty$. Prove that there exists arbitrarily large r, such that for every $0 < s < r$ $\int_{r-s}^{r+s} h \geqslant \int_{r-s}^{r+s} g$.

4.133 Prove for arbitrary continuous functions f and g on $[0, 1]$ the inequality $(\int_0^1 f)^2 + (\int_0^1 g)^2 \leqslant (\int_0^1 (f^2 + g^2)^{1/2})^2$. Give a geometric interpretation of this inequality.

4.134 Suppose that the nonnegative continuous functions a, b, c, f on $\mathbb{R}^+$ satisfy for all $x \geqslant 0$ $f(x) \leqslant \int_0^x [a(t)f(t) + b(t)] \, dt + c(x)$. Show the Gronwall inequality $f(x) \leqslant [\int_0^x b(t) \, dt + \max_{0 \leqslant t \leqslant x} c(t)] \exp(\int_0^x a(t) \, dt)$.

4.135 The functions $u, v, w : \mathbb{R}^+ \to \mathbb{R}^+$ are continuous, $u_0 > 0$, $0 \leqslant p < 1$, $q = 1 - p$. From the inequality $u(x) \leqslant u_0 + \int_0^x v(t)u(t) \, dt + \int_0^x w(t)u^p(t) \, dt$ deduce the estimate

$$u(x)\exp\left(-\int_0^x v(t)\,dt\right) \leqslant \left(u_0^q + q\int_0^x w(t)\exp\left(-q\int_0^t v(s)\,ds\right)dt\right)^{1/q}.$$

4.136 Prove the Nagumo theorem: If f is a nonnegative continuous function on $[0, 1]$, $f(0) = 0$, $\lim_{t \to 0} f(t)/t = 0$ and for all t $f(t) \leqslant \int_0^t f(s)s^{-1} \, ds$, then $f \equiv 0$.

4.137 Define for a continuous function $f : \mathbb{R}^+ \to \mathbb{R}^+$ such that $P(s) = \int_0^\infty f(u)u^s \, du < \infty$ for every $s > 0$ and for $a \in (0, 1)$ the function $g : \mathbb{R}^+ \to \mathbb{R}^+$ by $\int_0^{g(s)} f(u)u^s \, du = aP(s)$. Show that g increases.

4.138 Suppose that the function f defined on $\mathbb{R}^+$ is positive nonincreasing and convex, $f(0) = 0$ and the line $y = x - a$, $a > 0$, is its asymptote. Show that the functions f_n defined recurrently by $f_1 = f$, $f_n(x) = 2^{-1}f_{n-1}(f_{n-1}(2x))$ have the same properties and $f_n(x) \to \max(0, x - a)$ uniformly.

4.139 Find the lowest upper bound of $s(f) = \inf_{x > 0} \{f(x)x^{-1} \int_0^x (1 - f(t)) \, dt\}$, where f is a continuous function on $\mathbb{R}^+$.

4.140 Consider the function $f(x) = \sin(1/x)$ for $x \neq 0$ and $f(0) = 0$. Is f a derivative of some function F?

4.141 For which real a is the function $f(x) = \int_0^x |t|^a \sin(1/t)^{1/|t|} \, dt$ differentiable at the origin?

4.142 The functions f and g are differentiable in an open interval. Show that between two consecutive zeros of f there is a zero of $f' + fg'$.

4.143 Does there exist a differentiable function $f : \mathbb{R} \to \mathbb{R}$ such that $f(\mathbb{Q}) \subseteq \mathbb{Q}$ and $f'(\mathbb{Q}) \subseteq \mathbb{R} \backslash \mathbb{Q}$?

4.144 Suppose that for a function f, $f'(x) < 0 < f''(x)$ for $x < x_0$ and $f'(x) > 0 > f''(x)$ for $x > x_0$. Show that f is not differentiable at x_0.

4.145 Given a differentiable function f on (a, b) and arbitrary $x \in (a, b)$ does there exist $a < x_1 < x < x_2 < b$ such that $(f(x_2) - f(x_1))/(x_2 - x_1) = f'(x)$? Answer the same question supposing $f'' \neq 0$ in (a, b).

4.146 Given a differentiable function f on the interval (a, b) define the function $F : (a, b) \times (a, b) \to \mathbb{R}$ by the formula $F(x, y) = (f(x) - f(y))/(x - y)$ if $x \neq y$ and $F(x, x) = f'(x)$. Give necessary and sufficient conditions for
  (i) differentiability of F
  (ii) continuity of the derivative of F

4.147 For a twice differentiable function f on $\mathbb{R}$ such that f, f', and f'' increase and for $a \leqslant b$, $x > 0$, choose $\zeta = \zeta(x)$ satisfying
$$\frac{f(b + x) - f(a - x)}{b - a + 2x} = f'(\zeta).$$
Show that $\zeta(x)$ is increasing. Can one replace the condition "f'' increases" with "$f'' > 0$"?

4.148 Prove that for a real function f defined on an interval I the following conditions are equivalent

(i) $\left| \det \begin{bmatrix} f(u) & f(v) & f(w) \\ u & v & w \\ 1 & 1 & 1 \end{bmatrix} \right| \leqslant |(u - v)(v - w)(w - u)|$.

(ii) f is differentiable and f' satisfies the Lipschitz condition with the constant 2.

4.149 Define for a function f on $[a, b]$ with a decreasing continuous derivative, $f'(b) > 0$, the recurrent sequence $b_0 = b$, $b_n = f^{-1}(f(b_{n-1}) - f'(b_{n-1})(b_{n-1} - a))$. Show that $f(b_n)$ converges to $f(a)$.

4.150 Consider the iterates $x_0 = a$, $x_n = f(x_{n-1})$ for a differentiable function $f : [0, a] \to \mathbb{R}$ such that $0 < f(x) < x$, $f'(x) \geqslant c > 0$ and $f(x)/x$ is nonincreasing on $[0, a)$. Prove that
  (i) $\sum_{n=1}^{\infty} x_n < \infty$ if and only if $\int_0^a x/(x - f(x)) \, dx < \infty$.
  (ii) If $f'(0) < 1$ then $\sum_{n=1}^{\infty} x_n < \infty$.
  (iii) If $f'(0) = 1$ and f'' is bounded from below, then $\sum_{n=1}^{\infty} x_n = \infty$.
  (iv) $\sum_{n=1}^{\infty} x_n^s < \infty$ if and only if $\int_0^a x^s/(x - f(x)) \, dx < \infty$ $(s > 0)$.

4.151 The functions f, g, f_n, f', and f_n' are continuous in an open interval and $\lim_{n \to \infty} f_n(x) = f(x)$, $\lim_{n \to \infty} f_n'(x) = g(x)$. Show that $f' = g$.

4.152 Suppose that $f_n \in C^1[0, 1]$, $n = 1, 2, \ldots, f_n \to 0$ in $C[0, 1]$ but not in $C^1[0, 1]$ (for example, $f_n(x) = \sin(n\pi x)/n$). Show that there exists a function

$g \in C[0, 1] \backslash C^1[0, 1]$ which cannot be uniformly approximated by finite linear combinations of the functions f_n.

4.153 Consider a convex differentiable function f on $[0, 1/2]$, $f(0) = 0$, $f(1/2) = 1$, $f'(0) > 1$. Extend f to $[0, 1]$ putting $f(x) = f(1 - x)$. Prove that the set of points x such that the sequence of iterates $x_{n+1} = f(x_n)$, $x_0 = x$, is eventually periodic is dense in $[0, 1]$.

4.154 Suppose that the function f is nonnegative, convex and differentiable on the interval I and $x \in I$ is the unique point where f vanishes. Define the sequence (x_n) by $x_{n+1} = x_n - f(x_n)/f'(x_n)$ (the Newton iteration method). For every $x_0 \in I$ the sequence (x_n) tends to x. The convergence may be slow but prove that in any case $\sum_{n=1}^{\infty} (2^n f(x_n))^2 < \infty$.

4.155 Show that if $f \in C[0, 1]$, $\int_0^1 f = 0$, then $(\int_0^1 x^2 f(x) \, dx)^2 \leqslant \frac{1}{3} \int_0^1 x^2 (f(x))^2 \, dx$.

4.156 Show that for $f \in C^1[0, a]$ $|f(0)| \leqslant \frac{1}{a} \int_0^a |f| + \int_0^a |f'|$.

4.157 Prove for $f \in C^1[0, 1]$ such that $f(0) = f(1) = 0$ the inequality $|f(x)|^2 \leqslant \frac{1}{2}(\tanh \frac{1}{2}) \int_0^1 [|f(t)| + |f'(t)|]^2 \, dt$. Give an example showing that the constant above cannot be improved.

4.158 Show that if $f \in C^1[a, b]$, $f(a) = 0$ and $0 \leqslant f'(x) \leqslant 1$, then $\int_a^b (f(x))^3 \, dx \leqslant (\int_a^b f(x) \, dx)^2$.

4.159 Show that for every function $f \in C^1[0, 1]$ vanishing at $x = 0$ and $x = 1$ the inequality $(\int_0^1 f(x) \, dx)^2 \leqslant \frac{1}{12} \int_0^1 (f'(x))^2 \, dx$ holds. Determine all functions f such that the equality holds.

4.160 Prove the Wirtinger inequality: If f is an absolutely continuous function on $[0, 2\pi]$, $f(0) = f(2\pi)$, $\int_0^{2\pi} f = 0$, then $\int_0^{2\pi} (f(x))^2 \, dx \leqslant \int_0^{2\pi} (f'(x))^2 \, dx$. The equality holds if and only if $f(x) = A \cos x + B \sin x$.

Show a discrete analog of this inequality: For real numbers $x_1, \ldots, x_n$, $x_{n+1} = x_1$ such that $\sum_{k=1}^n x_k = 0$ the inequality $4(\sin \frac{\pi}{n})^2 \sum_{j=1}^n x_j^2 \leqslant \sum_{j=1}^n (x_{j+1} - x_j)^2$ is true with the equality only for $x_j = A \cos \frac{2\pi}{n}(j - 1) + B \sin \frac{2\pi}{n}(j - 1)$.

4.161 Prove that for every function $f \in C^2[a, b]$, $a \leqslant 0$, $b \geqslant 2$, the following inequality holds

$$\int_a^b (f''(x))^2 \, dx \geqslant \frac{3}{2} (f(0) - 2f(1) + f(2))^2$$

with the equality if and only if $f(x) = A + Bx + C(x_+^3 - 2(x - 1)_+^3 + (x - 2)_+^3)$, where $t_+ = \max(0, t)$.

4.162 Show that for $f \in C[a,b] \cap C^2(a,b)$, $f(a) = f(b) = 0$ $\int_a^b |f(x)| \, dx$
$\leqslant \dfrac{(b-a)^3}{12} \sup_{a<x<b} |f''(x)|$. Moreover if $f((a+b)/2) = 0$, then the constant $1/12$
can be replaced by $1/24$.

4.163 Let for a fixed real number a $K = \{f \in C^2[0,1] : f(0) = 0, \ f'(0) = a, \ f(1) = 0\}$. Find $\min_{f \in K} \{\int_0^1 (f''(x))^2 \, dx\}$ and a function which minimizes this functional.

4.164 Given $f \in C^2[0,1]$, $f(0) = 0$, determine

$$\lim_{n \to \infty} n^2 \left(\int_0^1 f(x) \, dx - \frac{2}{2n+1} \sum_{k=1}^{n} f\left(\frac{2k}{2n+1} \right) \right)$$

Apply this result to compute $\lim_{n \to \infty} n^2(\log 2 - \sum_{k=1}^{2n-1} 2/(2k+1))$.

4.165 Suppose that f is differentiable in a neighborhood of x and f' is continuous at x. Show that

$$\lim_{n \to \infty} \sum_{k=1}^{n} \left[f\left(x + \frac{k}{k^2 + n^2} \right) - f(x) \right] = \frac{\log 2}{2} f'(x)$$

4.166 Prove that any continuous function on the interval (a,b) satisfying for every x the condition $\lim_{h \to 0} h^{-3} \int_0^h [f(x+u) + f(x-u) - 2f(x)] \, du = 0$ must be linear.

4.167 Show that for every differentiable $f : [a,b] \to \mathbb{R}^+$, $f(a) = f(b) = 0$, there exists $c \in [a,b]$ such that $|f'(c)| \geqslant 4(b-a)^{-2} \int_a^b f(t) \, dt$.

4.168 Show that for $f \in C[0,1]$ satisfying $\int_0^1 x^k f(x) \, dx = 0$ for $k = 0, 1, \ldots, n-1$, $\int_0^1 x^n f(x) \, dx = 1$, there exists $x_0 \in [0,1]$ such that $|f(x_0)| \geqslant 2^n(n+1)$.

4.169 Prove that if $f : [a,b] \to \mathbb{R}$ satisfies the following conditions: $f(a) = f(b) = 0$, $f(x) > 0$ for $a < x < b$ and $f + f'' > 0$, then $b - a \geqslant \pi$.

4.170 Show that if $f : \mathbb{R} \to [-1,1]$ is twice continuously differentiable and $(f(0))^2 + (f'(0))^2 = 4$, then there exists x_0 such that $f(x_0) + f''(x_0) = 0$.

4.171 Show that $f \in C^2(\mathbb{R})$ satisfies for all x $(f(x))^2 \leqslant 1$, $(f'(x))^2 + (f''(x))^2 \leqslant 1$, then for all x $(f(x))^2 + (f'(x))^2 \leqslant 1$.

4.172 Show that if $f \in C^2(\mathbb{R}^+)$ and the integrals $\int_0^\infty f$, $\int_0^\infty |f''|$ converge, then $\lim_{x \to \infty} f(x) = \lim_{x \to \infty} f'(x) = 0$.

4.173 Suppose that $f : \mathbb{R}^+ \to \mathbb{R}$ satisfies $\lim_{x \to \infty} (f + f')(x) = 0$. Show that $\lim_{x \to \infty} f(x) = 0$.

4.174 Prove that if $f \in C^2(\mathbb{R}^+)$ and $\lim_{x \to \infty} (f + f' + f'')(x) = C$, then $\lim_{x \to \infty} f(x) = C$.

4.175 Prove the following version of the Hardy-Littlewood theorem: If f is twice differentiable on $\mathbb{R}^+$, $f(x) = o(1)$, $f''(x) = O(1)$ for $x \to \infty$, then $f'(x) = o(1)$.

Remark: the assumption $f''(x) = O(1)$ can be replaced by the condition $\int_x^{x+1} s(|f''(u)|)\, du = O(1)$, where s is a positive increasing convex function such that $s^{-1}(x) = o(x)$. This last condition cannot be replaced by $s^{-1}(x) = O(x)$.

4.176 Prove the Littlewood theorem: If a function $f: \mathbb{R}^+ \to \mathbb{R}$ is $n+1$ times differentiable, $\lim_{x \to \infty} f(x)$ is finite, and $f^{(n+1)}$ is bounded, then $\lim_{x \to \infty} f^{(n)}(x) = 0$.

4.177 Consider the function $f(t) = ((1 - c_1 t) \cdots (1 - c_n t))^{-1}$, where $c_1, \ldots, c_n \in \mathbb{R}$ satisfy $c_1^k + \cdots + c_n^k > 0$ for every natural k. Show that all the derivatives of f at the origin are positive.

4.178 Denote by δ the differential operator $x\, d/dx$. Show that if P is a polynomial, then $e^{\delta} P(x) = P(ex)$.

4.179 Verify that for the functions f_n defined recurrently by $f_0 \equiv 1$,

$$f_{n+1}(x) = x^{f_n(x)} \frac{df_n}{dx}(x) = \frac{1}{x} \Sigma_{j=0}^{n-1} (\log x)^j \Pi_{i=n-j-2}^{n-1} x^{f_i(x)}.$$

4.180 Find all differential operators with smooth coefficients $A = \Sigma_{k=0}^n a_k(x) d^k/dx^k$, $B = \Sigma_{k=0}^m b_k(x) d^k/dx^k$ such that $A \circ d/dx = B \circ x$.

4.181 Show that if f is a continuous function on $\mathbb{R}$ such that $\int_{\mathbb{R}} fg'' \geq 0$ for every nonnegative twice differentiable function g, then f is convex.

4.182 Prove the du Bois-Reymond lemma: If a and b are continuous functions on $[0, 1]$ and for every $g \in C^1[0, 1]$ such that $g(0) = g(1) = 0$ $\int_0^1 (ag' + bg) = 0$, then a is continuously differentiable and $b = a'$.

4.183 Let the continuous functions $f, g: \mathbb{R}^+ \to \mathbb{R}$ be equal identically to 1 on $[0, 1]$, f, g being differentiable on $(1, \infty)$ and satisfy for $x > 1$ the equations $xf'(x) = -f(x - 1)$, $xg'(x) = g(x - 1)$. Show that for $x \geq 0$ $\int_0^x f(x - t)g(t)\, dt = x$.

4.184 Show that if y is a solution of the differential equation $y'' + q(x)y = 0$, where q is a continuous function and $\int_0^\infty t|q(t)|\, dt < \infty$, then y' has a limit at $+\infty$.

4.185 Define for a continuous function $f: (0, 1) \to \mathbb{R}$ such that $\int_0^1 x^{-n-1}|f(x)|\, dx < \infty$ $y(x) = \int_0^1 G(x, t)f(t)\, dt$, where $G(x, t) = x^n(t^n - t^{-n})/2n$ for $0 \leq x \leq t \leq 1$, and $G(x, t) = t^n(x^n - x^{-n})/2n$ for $0 \leq t < x \leq 1$. Find the functions $p, q: (0, 1) \to \mathbb{R}$ such that $\frac{d}{dx}\left(p(x)\frac{dy}{dx}\right) + q(x)y = f(x)$.

4.186 Let $M_j = \sup_{t \in I} |f^{(j)}(t)|$, $j = 0, 1, 2$, for a twice differentiable function $f : I \to \mathbb{R}$. Prove that

  (i) $|f'(x)| \leqslant M_0/c + (x^2 + c^2)M_2/2c$, if $I = [-c, c]$.

  (ii) $M_1 \leqslant 2(M_0 M_2)^{1/2}$, if the length of I is $\geqslant 2(M_0/M_2)^{1/2}$.

  (iii) $M_1 \leqslant (2M_0 M_2)^{1/2}$, if $I = \mathbb{R}$.

Give the examples showing optimality of the constants in (ii) and (iii).

Deduce that for p times differentiable function $M_k \leqslant 2^{k(p-k)/2} M_0^{1-k/p} M_p^{k/p}$, $k = 1, 2, \ldots, p$.

4.187 Let $f : [-1, 1] \to [-1, 1]$ be a smooth function, $J = [x_0, x_0 + d] \subseteq [-1, 1]$ and $a + b + c = d$ for some $a, b, c > 0$. Prove that for every natural k

$$\min_{x \in J} |f^{(k)}(x)| \leqslant \frac{1}{b} \left(\min_{x \in [x_0, x_0 + a]} |f^{(k-1)}(x)| + \min_{x \in [x_0 + a + b, x_0 + d]} |f^{(k-1)}(x)| \right)$$

and

$$\min_{x \in J} |f^{(k)}(x)| \leqslant 2^{k(k+1)/2} k^k / d^k.$$

4.188 Prove the Faà di Bruno formula:

$$D^n(f \circ g)(x) = \sum \frac{n!}{k_1! \cdots k_n!} (D^k f)(g(x)) \left(\frac{Dg(x)}{1!} \right)^{k_1} \cdots \left(\frac{D^n g(x)}{n!} \right)^{k_n},$$

where the sum is taken over all $k_1, \ldots, k_n$ such that $k_1 + 2k_2 + \cdots + nk_n = n$, $k = k_1 + \cdots + k_n$.

4.189 Prove the following version of the Taylor formula: If $f \in C^2[a, b]$ is three times differentiable in (a, b), then there exists $c \in (a, b)$ such that

$$\begin{vmatrix} f(b) & b^2 & b & 1 \\ f(a) & a^2 & a & 1 \\ f'(a) & 2a & 1 & 0 \\ f''(a) & 2 & 0 & 0 \end{vmatrix} = -f'''(c) \frac{(b-a)^3}{3}.$$

Generalize to the case of n derivatives.

4.190 Consider the Taylor formula for a function f in a neighborhood of a with the rest in Lagrange form

$$f(x) = f(a) + f'(a)(x-a) + \cdots + \frac{f^{(n-1)}(a)}{(n-1)!}(x-a)^{n-1} + \frac{f^{(n)}(c)}{n!}(x-a)^n.$$

Suppose that for $j = 1, 2, \ldots, p-1$ $f^{(n+j)}(a) = 0$ and $f^{(n+p)}(a) \neq 0$. Show that

$$\lim_{x \to a} \frac{c-a}{x-a} = \left(\frac{n! \, p!}{(n+p)!} \right)^{1/p}.$$

4.191 Let P be a polynomial with all the roots of negative real part. Show that if the function f satisfies $\lim_{x\to\infty} P(d/dx)f(x) = 0$, then $\lim_{x\to\infty} f(x) = 0$.

4.192 Suppose that a function $f \in C^n[a,b]$ has at least $n+1$ zeros and all the roots of the polynomial $P(x) = x^n + c_{n-1}x^{n-1} + \cdots + c_0$ are real. Show that the function $P(d/dx)f$ has a zero in $[a,b]$.

4.193 Show that if $f \in C^n[0,1]$ and $f^{(k)}(0) = f^{(k)}(1) = 0$ for $k = 0, 1, \ldots, n-1$, then $\int_0^1 |f^{(n)}|^p \geq (2n+1)^{-\max(1,p/2)} \left(\dfrac{(2n+1)!}{n!}\right)^p \left|\int_0^1 f\right|^p$.

4.194 Show that if P is a polynomial of degree n such that $\int_0^1 x^k P(x)\, dx = 0$ for $k = 1, \ldots, n$, then $\int_0^1 (P(x))^2\, dx = (n+1)^2 (\int_0^1 P(x)\, dx)^2$.

4.195 Prove that for $f \in C^n[0,1]$ satisfying $f^{(j)}(0) = f^{(j)}(1) = 0$ for $j = 1, \ldots, n-1$ and $\int_0^1 x^k f(x)\, dx = 0$ for $k = 1, \ldots, m$ the inequality $(\int_0^1 f)^2 \leq (2n+1)\left(\dfrac{n!\,m!}{(2n+m+1)!}\right)^2 \int_0^1 (f^{(n)})^2$ holds.

4.196 Prove that $f \in C_0^\infty(\mathbb{R})$ such that $\int_{\mathbb{R}} f(x)\, dx = \int_{\mathbb{R}} x f(x)\, dx = 0$ can be represented as $f = g''$ for some $g \in C_0^\infty(\mathbb{R})$.

4.197 Show that if a smooth function f satisfies the estimate $f(x) = o(x^n)$ as $x \to \pm\infty$, then $f^{(r)}$ has a zero for every $r \geq n+1$.

4.198 Let f belong to $C^\infty(\mathbb{R}^+)$, $f(0)f'(0) \geq 0$ and $\lim_{x\to\infty} f(x) = 0$. Show that there exists a strictly increasing sequence of reals (x_n) such that $f^{(n)}(x_n) = 0$ for every n.

4.199 Prove that if f is a smooth function on $[0,1]$ and for every $x \in [0,1]$ there exists $n = n(x)$ such that $f^{(n)}(x) = 0$, then f is a polynomial. Is this conclusion true assuming only $\lim_{n\to\infty} f^{(n)}(x) = 0$?

4.200 Consider an increasing diffeomorphism f of $\mathbb{R}$ such that $f(0) = 0$, $f'(0) = a$ and $|f(x) - f(y)| \leq k|x - y|$ for some $0 < k < 1$ and all x, y. Prove that there exists a diffeomorphism h such that $h \circ f \circ h^{-1}(x) = ax$.

4.201 The functions f, $g : \mathbb{R} \to \mathbb{R}$ are said to be similar if there exists a bijection h such that $f = h^{-1} \circ g \circ h$. Are the functions $\sin$ and $\cos$ similar? Answer the same question for x^2 and $x^2 + ax + b$.

4.202 Define the sequence of functions on $\mathbb{R}$ by $f_0(x) = \operatorname{sgn} x$, $f_n(x) = h_n^{-1} \int_{x-h_n}^x f_{n-1}(t)\, dt$ for $n \geq 1$, where $h_n > 0$, $\sum_{n=1}^\infty h_n < \infty$. Prove that the sequence (f_n) converges to a smooth function f such that $f(x) = -1$ for $x < 0$, $f(x) = 1$ for $x > \sum_{n=1}^\infty h_n$ and $|f^{(n)}(x)| \leq 2^n (h_1 \cdots h_n)^{-1}$.

4.203 Construct a smooth function positive on the set of rationals and having uncountably many zeros.

4.204 Is the set $\{x : \operatorname{sgn} f$ is not continuous at $x\}$ of measure zero for every smooth function f?

4.205 Construct a smooth function on $(0, 1)$ which is not analytic on any open interval.

4.206 Is every smooth function on $\mathbb{R}$ supported on $[0, 1]$ representable as a difference of two nonnegative smooth functions supported in $[0, 1]$?

4.207 Suppose that $f \in C^\infty[0, 1]$ satisfies: $f \not\equiv 0$, $f^{(k)}(0) = 0$ for every k and $\sum_{k=1}^\infty a_k f^{(k)}(x)$ converges uniformly on $[0, 1]$. Show that $\lim_{k \to \infty} k! a_k = 0$.

4.208 Consider a smooth function f on $(-1, 1)$ such that $f(0) = 0$. Show that the function $g : (-1, 1) \backslash \{0\} \to \mathbb{R}$ defined by $g(x) = f(x)/x$ can be extended to a smooth function on $\mathbb{R}$ and $\lim_{x \to 0} g^{(k)}(x) = (k + 1)^{-1} f^{(k+1)}(0)$.

4.209 If $f : \mathbb{R} \to \mathbb{R}$ is an even function then $f(x) = g(x^2)$ for some $g : \mathbb{R}^+ \to \mathbb{R}$. Prove that
   (i) If f is in C^{2k}, then g is in C^k.
   (ii) If f is analytic, then g is also analytic.
Give an analog for odd functions.

4.210 (c_n) is an arbitrary sequence of reals. Prove that there exists a strictly decreasing to zero sequence (a_n) and a sequence of smooth functions on $\mathbb{R}(u_n)$ such that $|u_n^{(k)}(t)| \leqslant 2^{-n}$ for $|t| \leqslant a_{n+1}$, $k = 0, 1, \ldots, n - 1$, $u_n^{(k)}(t) = 0$ for $|t| \leqslant a_{n+1}$, $k \geqslant n + 1$, and $u_n^{(n)}(0) = c_n$.
   Prove the Borel theorem: There exists a smooth function f on $\mathbb{R}$ such that $f^{(n)}(0) = c_n$ for every n.
   Generalize this theorem to the case $f : \mathbb{R}^m \to \mathbb{R}$.

4.211 $f : \mathbb{R} \to \mathbb{R}$ is a function such that the powers f^2 and f^3 are smooth. Is f smooth?

4.212 Prove that if for every $t \in (a, b)$ the radius of convergence of the Taylor series of a smooth function f in t is at least equal to $c > 0$, then f is analytic.

4.213 Prove the Bernstein theorem: If a smooth function on $[0, r)$ satisfies the condition $f^{(n)}(x) \geqslant 0$ for every x and n, then its Taylor series converges to f (and f has analytic continuation into $\{z \in \mathbb{C} : |z| < r\}$).

4.214 Suppose that for a smooth function f on the interval $I \subseteq \mathbb{R}$ there exists $p \geqslant 0$ such that for every n $f^{(n)}$ has at most p zeros. Show that f is analytic on I.

4.215 Does there exist $x_0 > 0$ such that all the derivatives of the function $f(x) = x^x$ are nonnegative on $[x_0, \infty)$?

4.216 $f:(0, \infty) \to \mathbb{R}$ is said to be completely monotone if $(-1)^n f^{(n)}(x) \geq 0$ for all $x \in (0, \infty)$ and $n = 0, 1, \ldots$.

Prove the Bernstein theorem: Each completely monotone function is of the form $f(x) = \int_0^\infty e^{-ax} \, dm(a)$, where m is a unique Borel measure on $[0, \infty)$ (and $\lim_{x \to \infty} f(x) = m([0, \infty))$).

4.217 Construct a smooth function $f:\mathbb{R} \to \mathbb{R}$ satisfying the following conditions:

(i) Every solution of the differential equation $y' = f(y)$ is smooth.

(ii) For every function $h:\mathbb{R}^+ \to \mathbb{R}^+$ there exists a solution y of this equation such that $y(x) > h(x)$ for all $x \in \mathbb{R}^+$.

4.218 Let $f:\mathbb{R} \to \mathbb{R}$ be a continuous strictly increasing function and $g = f \circ f$ belongs to (i) $C^1(\mathbb{R})$, (ii) $C^\infty(\mathbb{R})$, (iii) $C^\omega(\mathbb{R})$. Is it true that f has the same regularity properties?

4.219 Suppose that the sequences of reals (a_n) and (b_n) tend to zero and $a_n \neq b_n$ for every n. Let f satisfies the following condition: for every $x \in (0, 1)$ there exists $N = N(x)$ such that for all $n \geq N$ $f(x + a_n) = f(x + b_n)$.

Is that function f necessarily constant if $0 < b_n < a_n$?

The same question for $b_n = -a_n$.

4.220 The functions $f_n:\mathbb{R} \to \mathbb{R}$ satisfy the Lipschitz condition with the constant L, $f:\mathbb{R} \to \mathbb{R}$ is continuous and for every x there exists n such that $f(x) = f_n(x)$. Show that f is also Lipschitzian with the same constant L.

Does the same hold for the Hölder condition?

5

Functional Equations and Functions of Several Variables

Topics covered include:
general functional equations, extremal values, multiple integrals, structure of differentiable mappings, integral transforms, special differential equations, harmonic functions, and geometric problems.

5.1 Find all continuous functions f such that $f(2x + 1) = f(x)$ for all real x.

5.2 Solve the functional equation $f(x) = f(x/(1 - x))$, $x \neq 1$, assuming that f is continuous at zero.

5.3 Determine all continuous functions f such that for every geometric progression (x_n) the sequence $(f(x_n))$ is also a geometric progression.

5.4 $C\text{-lim}_{n \to \infty} x_n = \lim_{n \to \infty} (x_1 + \cdots + x_n)/n$ denotes the limit of the sequence (x_n) in the Cesàro sense (if it exists). Find all real functions f which are Cesàro continuous, that is, $f(C\text{-lim } x_n) = C\text{-lim } f(x_n)$ for every Cesàro convergent sequence (x_n).

5.5 $f(x) = 1 - x$. Is it the only decreasing continuous transformation of $[0, 1]$ such that $f(f(x)) = x$?

5.6 Find all functions f satisfying the identity

$$f(x) + f((x - 1)/x) = 1 + x \quad \text{for} \quad x \neq 0, 1$$

5.7 Determine all functions f such that

$$f(x) + f(1/(1 - x)) = x \quad \text{for} \quad x \neq 1$$

5.8 Find all polynomials f such that

$$f(x^2) + f(x)f(x + 1) = 0$$

5.9 For which continuous real functions f is $f(x) - f(y)$ rational whenever $x - y$ is rational?

5.10 Find all homeomorphisms of $[0, 1]$ such that $f(2x - f(x)) = x$.

5.11 Determine all continuous real functions f such that $f(2x - f(x)/m) = mx$, where $m \neq 0$ is a fixed real number.

5.12 Determine the continuous solutions $f:(0, 1) \to \mathbb{R}$ of the functional equation

$$f((x - y)/(\log x - \log y)) = (f(x) + f(y))/2, \quad x \neq y$$

5.13 Compute $f(\sqrt{q})$ for positive rational q if f is an additive function satisfying $f(1/x) = f(x)/x^2$ for $x \neq 0$ and $f(1) = 1$.

5.14 Prove that the identity is the unique nonconstant additive and multiplicative function $f:\mathbb{R} \to \mathbb{R}$. Determine $f:\mathbb{C} \to \mathbb{C}$ with the same properties.

5.15 Let $f:\mathbb{R} \to \mathbb{R}$ be an additive function satisfying one of the conditions
  (i) f is bounded from above on an interval (a, b).
  (ii) f is locally integrable.
  (iii) f is Lebesgue measurable.
Show that $f(x) = f(1)x$.

5.16 An additive function $f:\mathbb{R} \to \mathbb{R}$ satisfies $f(x)f(1/x) > 0$ for $x \neq 0$. Must it be continuous? Answer this question for $f:\mathbb{R} \to \mathbb{C}$.

5.17 Determine all continuous solutions $f:\mathbb{R} \to \mathbb{R}$ of the d'Alembert functional equation

$$2f(x)f(y) = f(x + y) + f(x - y)$$

5.18 Find all continuous real solutions defined on an open interval of the functional equation

$$f((x + y)/(1 + xy)) = f(x) + f(y)$$

5.19 Solve the functional equation $\int_{x-y}^{x+y} f = f(x)f(y)$ in the class of locally integrable real functions.

5.20 Determine continuous real solutions of the functional equation

$$f(x + y) = f(x) + f(y) + kxy$$

5.21 For a given continuous strictly monotone function f on $\mathbb{R}^+$ define the mean M_f of the positive numbers $x_1, \ldots, x_n$ by the formula

$$M_f(x_1, \ldots, x_n) = f^{-1}\left(\frac{1}{n}\sum_{j=1}^{n} f(x_j)\right)$$

Show that the arithmetic mean M_x and the means $M_{\exp(bx)}$ (considered in information theory) are the only translation invariant means M_f, that is,

$$M_f(x_1 + a, \ldots, x_n + a) = M_f(x_1, \ldots, x_n) + a$$

Show that the power means M_{x^p} and the geometric mean $M_{\log x}$ are the only homogeneous means, that is, $M_f(cx_1, \ldots, cx_n) = cM_f(x_1, \ldots, x_n)$.

5.22 Prove that the only injective continuous functions $f: \mathbb{R}^+ \to \mathbb{R}^+$ defining a norm in $\mathbb{R}^n$ by the formula

$$\|u\| = f^{-1}\left(\sum_{j=1}^{n} f(|u_j|)\right) \qquad u = (u_1, \ldots, u_n)$$

are of the form $f(u) = u^p$ for $p \geqslant 1$.

5.23 Find all continuous functions $F: \mathbb{R}^2 \to \mathbb{R}$ satisfying the functional equation $F(x, y) + F(y, z) + F(z, x) = 3F((x + y + z)/3, (x + y + z)/3)$. What can be said about F without assuming continuity?

5.24 Prove the following converse of the Pythagorean theorem: if $F: \mathbb{R}^2 \to \mathbb{R}^+$ satisfies the condition $F(u + v) = F(u) + F(v)$ for all orthogonal vectors u, v, then $F(v) = c|v|^2$ for some $c \geqslant 0$.

5.25 For every fixed $n \geqslant 2$ find a function $f: \mathbb{R} \to \mathbb{R}$ such that for every x, $f(x) + f(2x) + \cdots + f(nx) = 0$ and $f(x) = 0$ if and only if $x = 0$.

5.26 Find general solution of the functional equation $\sum_{k=1}^{n} f(x^k) = \sum_{k=1}^{n} f(x^{-k})$, $x \neq 0$.

5.27 Does there exist a positive continuous function f on $[1, \infty)$ such that $\sum_{n=1}^{\infty} f(n) = \infty$ and $\sum_{n=1}^{\infty} a^n f(a^n) < \infty$ for all $a > 1$?

5.28 Determine all continuously differentiable functions f satisfying the identity $\int_0^1 f(tx)\,dt = cf(x)$ for a fixed real c.

5.29 Find all continuous solutions $f: \mathbb{R} \to \mathbb{R}$ of the functional equation $f(xyz) = f(x) + f(y) + f(z)$ in the case if

(i) x, y, z are arbitrary nonzero real numbers.

(ii) $a < x$, y, $z < b$ and $1 < a^3 < b$.

How can one weaken the continuity hypothesis? Extend the result for the functional equation

$$f(x_1 \cdots x_n) = f(x_1) + \cdots + f(x_n)$$

5.30 Check that $f_0(x) = \sum_{n=0}^{\infty} x^{2n}/(2n + 1)$ solves the functional equation $f(2x/(1 + x^2)) = (1 + x^2)f(x)$, $0 \leqslant x < 1$. Express f_0 in terms of elementary functions. Find general continuous solution of this equation.

5.31 Show that the function $\sin z$ satisfies the functional equation $|f(z)| = |f(\operatorname{Re} z) + f(i \operatorname{Im} z)|$. Find other analytic solutions of this equation.

5.32 Find all analytic solutions of the functional equation $|f(z)|^2 = |f(\operatorname{Re} z)|^2 + |f(i \operatorname{Im} z)|^2$.

5.33 Show that the function $|\tan z|^2$, $z = x + iy$, satisfies the equation $f(z) = (f(x) + f(y))/(1 + f(x)f(y))$.

5.34 Construct a real function satisfying the condition $\int_x^{x^2} f(t)\, dt = 1$ for every $x > 1$.

5.35 Show that for $c > \frac{1}{2}$ there is no function $f : [0, 1] \to \mathbb{R}$ satisfying the equation

$$f(x) = 1 + c \int_0^1 f(y)f(y - x)\, dy$$

5.36 Solve the functional equation $(f(x) - f(y))/(x - y) = f'(px + qy)$, where x, y are real and p, q are fixed positive numbers with $p + q = 1$.

5.37 Determine all twice differentiable functions such that for every $x > 0$ the length of the graph of f on $[0, x]$ is equal to the area of the set between this graph and the x-axis.

5.38 Prove that the functional equation

$$F(x_1, \ldots, x_n) = 1 + \int_0^{x_1} \cdots \int_0^{x_n} F(y_1, \ldots, y_n)\, dy_1 \cdots dy_n$$

has a unique solution.

5.39 Find all solutions of the functional differential equation $f' = f^{-1}$ such that $f(0) = 0$ and $f'(x) > 0$ for $x > 0$.

5.40 Solve the differential equation $\delta(\delta - 1)(\delta - 2) \cdots (\delta - n + 1)y(x) = g(x)$, where $\delta = x\, d/dx$, g is a given continuous function, and $y(1) = y'(1) = \cdots = y^{(n-1)}(1) = 0$.

5.41 Find all rational solutions $y: \mathbb{R} \to \mathbb{R}$ of the Riccati equation $y' = y^2 + a/x - b$, $a, b \neq 0$.

5.42 Prove that the Bernoulli equation $y' + y^2 + xy = 0$ has exactly two solutions with the nonvanishing second derivative on $\mathbb{R}$. Then show that the equation $y' + y^2 + x^{2n-1}y = 0$ has at most two such solutions.

5.43 Determine analytic solutions of the differential equation $xy'' - (x + n)y' + ny = 0$, $n = 1, 2, 3, \ldots$.

5.44 Find differentiable solutions $f: \mathbb{R} \to [1, \infty)$ of the equation

$$\int_1^{f(x)} \exp(u^2)\, du = \int_0^x \frac{u}{f(u)}\, du \qquad x \in \mathbb{R}$$

5.45 Solve the equation

$$\frac{dy(t)}{dt} = y(t) + \int_0^1 y(s)\, ds \qquad t \geq 0,\ y(0) = 1$$

5.46 Suppose that $f: \mathbb{R}^+ \to \mathbb{R}$ satisfies the conditions $f'(x) = f(cx)$ for some $c > 1$ and $f(0) = 0$. Must it be constant?

5.47 Construct a function $f: [0, \infty) \to \mathbb{R}$ satisfying the following conditions:

  (i) $f'(x) = 2f(2x)$ for $x \geq 0$

  (ii) $f(1 + x) = f(1 - x) = 1 - f(x)$ for $0 \leq x \leq 1$

  (iii) $f(x)$ does not tend to 0   if $x \to \infty$

5.48 Show that a measurable function $f: \mathbb{R} \to \mathbb{C}$ of polynomial growth which satisfies the equation $f(x) = \frac{1}{2}\int_{x-1}^{x+1} f(t)\, dt$ must be linear. Is the growth condition imposed on f essential? Consider multidimensional generalization.

5.49 Determine all continuous functions $f: \mathbb{R} \to \mathbb{R}$ such that for some continuous function $g: \mathbb{R}^2 \to \mathbb{R}$ $f(xy) = g(x, f(y))$ for every real x, y.

5.50 Prove that a function f defined for all real x, y, which is a polynomial in x for every fixed y and vice versa, must be a polynomial in two variables.
  Is it true for functions defined for rational x, y only?

5.51 Given a continuous function $f: \mathbb{R}^n \to \mathbb{R}$, denote by X_f the subspace of $C(\mathbb{R}^n)$ spanned by the translates of f ($T_a f(x) = f(x - a)$, $a \in \mathbb{R}^n$) and the dilations of f ($I_c f(x) = f(cx)$, $c \in \mathbb{R}$).
  Prove that f is a polynomial if and only if X_f is finite dimensional.

5.52 Prove the Steinitz theorem on the conditionally convergent vector series: If $(x_k) \subset \mathbb{R}^n$, $\Sigma_{k=1}^\infty x_k = 0$, $\Sigma_{k=1}^\infty |x_k| = \infty$, then the set F of sums of all rearrangements of the series $\Sigma\, x_k$ is the orthogonal complement of its domain of absolute convergence, that is, $F = \{x^* \in (\mathbb{R}^n)^* : \Sigma_{k=1}^{\infty} |x^*(x_k)| < \infty\}^{\perp}$.

Draw as corollaries the Riemann $(n = 1)$ and the Lévy $(n = 2, \mathbb{R}^2 = \mathbb{C})$ theorems.

5.53 Given a continuous function $f : [0, 1] \times [0, 1] \to \mathbb{R}$ let $y^*(a) = \inf\{y : f(a, y)$ attains its maximal value$\}$.

Is the function y^* continuous? Is it continuous if $f(a, y)$ attains its maximum at only one point for every a?

5.54 Let nonnegative continuous functions f and g on the parallelepiped $P = \Pi_{j=1}^{n}[0, a_j]$ satisfy the conditions

(i) $f(x_1, \ldots, x_n) = g(x_1, \ldots, x_n) = 0$ if $\Pi_{j=1}^{n} x_j = 0$,

(ii) $f(x_1, \ldots, x_n) > 0$ if $\Pi_{j=1}^{n} x_j > 0$.

Show that there exists a strictly increasing continuous function $h : \mathbb{R}^+ \to \mathbb{R}$ such that $h(0) = 0$ and $h \circ g(x_1, \ldots, x_n) < f(x_1, \ldots, x_n)$ if $\Pi_{j=1}^{n} x_j > 0$.

5.55 Find necessary and sufficient conditions for the function $x^a y^b$ to be linearly bounded on $\mathbb{R}^+ \times \mathbb{R}^+$, that is, to satisfy the inequality $x^a y^b \leqslant C(x + y)$ for some $C > 0$.

5.56 P is a homogeneous real polynomial in n variables, $D = P(\partial/\partial x_1, \ldots, \partial/\partial x_n)$. Show that any nonzero polynomial solution u of the equation $Du = 0$ is not divisible by P.

5.57 Characterize all continuous functions f on the unit circle satisfying $f(z^2) = f(z)^2$.

5.58 The function f is continuous on the closed unit disk and injective in its interior. Show that the set of points on the boundary for which $f^{-1}(\{x\})$ contains at least three elements is countable.

5.59 Determine fixed and periodic points of the plane transformation $T(x, y) = (x + y, y - (x + y)^3)$.

5.60 Prove the inequality

$$|a - c|\,|b - d| \leqslant |a - b|\,|c - d| + |a - d|\,|b - c|$$

for every $a, b, c, d \in \mathbb{R}^n$.

5.61 Find the minimum of the function

$$(r - 1)^2 + (s/r - 1)^2 + (t/s - 1)^2 + (4/t - 1)^2$$

for real r, s, t such that $1 \leqslant r \leqslant s \leqslant t \leqslant 4$.

5.62 Determine the extremal values of the function $f(x, y, z) = x^2 + y^2 + z^2$ on the surface $x^4/a^4 + y^4/b^4 + z^4/c^4 = 1$.

5.63 Let for $x = (x_1, \ldots, x_n)$, $j = 1, \ldots, n$,

$$S_j(x) = \sum_{1 \leqslant k_1 < \cdots < k_j \leqslant n} x_{k_1} \ldots x_{k_j}$$

denote the jth elementary symmetric function of the variables $x_1, \ldots, x_n$. Find the extremal values of $S_n(x)$ subject to the constraints $S_j(x) = s_j$, $j = 1, \ldots, n-1$.

5.64 Compute the extremal values of the p-power means, $p > 0$, of the lengths of the sides of the triangle inscribed in the unit circle.

5.65 Find

$$\lim_{x \to \infty} e^{-x} \int_0^x \int_0^x \frac{e^u - e^v}{u - v}\, du dv$$

5.66 Compute

$$\lim_{x \to \infty} x^4 \exp(-x^3) \int_0^x \int_0^{x-u} \exp(u^3 + v^3)\, dv du$$

5.67 Examine convergence of the integrals

$$I_n(a) = n^a \int_D (1 - x - y)^n f(x, y)\, dxdy$$

where $f \in C^2(\mathbb{R} \times \mathbb{R})$, $a \in \mathbb{R}$, $D = \{(x, y) : x^2 + y^2 \leqslant 1, x, y \geqslant 0\}$.

5.68 Show that for every $c > 0$

$$\int_1^\infty \cdots \int_1^\infty \frac{dx_1 \cdots dx_n}{x_1 \cdots x_n (\max(x_1, \ldots, x_n))^c} < \infty$$

5.69 Prove that if $f \in L^p(\mathbb{R})$ and $p = (n+1)/n$, then

$$\int \cdots \int |f(x_1) \cdots f(x_n) f(x_1 + \cdots + x_n)|\, dx_1 \cdots dx_n \leqslant \|f\|_p^{n+1}$$

5.70 Prove convergence of the integral

$$I = \int \cdots \int f(y_1) \cdots f(y_n) f(y_1 + \cdots + y_n)\, dy_1 \cdots dy_n$$

where $f(y) = \min(1, 1/|y|)$.

Deduce absolute convergence of the integral

$$\int \cdots \int \frac{\sin y_1}{y_1} \cdots \frac{\sin y_n}{y_n} \frac{\sin(y_1 + \cdots + y_n)}{y_1 + \cdots + y_n}\, dy_1 \cdots dy_n$$

Show that for $a_j > 0$

$$\int \cdots \int \frac{\sin a_1 x_1}{x_1} \cdots \frac{\sin a_n x_n}{x_n} \frac{\sin(a_1 x_1 + \cdots + a_n x_n)}{x_1 + \cdots + x_n} dx_1 \cdots dx_n$$

$$= \pi^n \min(a_0, \ldots, a_n)$$

5.71 Compute

$$\lim_{n \to \infty} n^{-1/2} \int_0^1 \cdots \int_0^1 (x_1^2 + \cdots + x_n^2)^{1/2} dx_1 \cdots dx_n$$

5.72 Show that for any $N \times N$ positive definite matrix A and a symmetric matrix B

$$\int_{\mathbb{R}^N} \exp(-(x, Ax) - i(x, Bx)) \, dx = \left(\frac{\pi^N}{\det(A + iB)} \right)^{1/2}$$

5.73 Compute for $f \in C([0, 1] \times [0, 1])$

$$\lim_{n \to \infty} \left(\frac{(2n + 1)!}{(n!)^2} \right)^2 \int_0^1 \int_0^1 (xy(1 - x)(1 - y))^n f(x, y) \, dx \, dy$$

5.74 Find the limit

$$\lim_{n \to \infty} n \left(\int_0^1 \int_0^1 f(x, y) \, dx \, dy - n^{-2} \sum_{j,k=1}^n f(j/n, k/n) \right)$$

for $f \in C^1([0, 1] \times [0, 1])$.

5.75 Determine $\lim_{t \to \infty} \int_a^b f(x, \sin tx) \, dx$ for functions $f \in C([a, b] \times [-1, 1])$.

5.76 A function f is continuous and positive on $\mathbb{R}$, $\int_{-\infty}^\infty f = 1$. For fixed $r > 0$ find the limit

$$\lim_{n \to \infty} \int \cdots \int_{\{\sum_{k=1}^n x_k^2 \le r^2\}} \prod_{k=1}^n f(x_k) \, dx_k$$

5.77 Examine continuity and differentiability of the function f defined by $f(x, y) = |x|^a y/(x^2 + y^4)$ for $x \ne 0$, and $f(0, y) = 0$.

5.78 Study the differentiability of the function given by

$$f(x, y) = \begin{cases} (xy)^{1/2} \sin(x^2 + y^2)^{-1/2} & \text{for} \quad (x, y) \ne (0, 0) \\ 0 & \text{for} \quad (x, y) = (0, 0). \end{cases}$$

5.79 All partial derivatives $\partial^{m+n} f / \partial x^n \partial y^m$ of a function $f : \mathbb{R}^2 \to \mathbb{R}$ exist everywhere. Does it imply continuity of f?

5.80 The exam question was as follows: if $xy = 4$, find d^2y/dx^2 and d^2x/dy^2. The student found d^2y/dx^2 correctly. He then wrote down its reciprocal and labeled that d^2x/dy^2—and it was right. The professor claimed that this was just luck and that it could not happen with any other function. Was the professor right?

5.81 Show that if a real function f is differentiable on an open set containing the unit ball B in $\mathbb{R}^n$ and $|f(x)| \leqslant 1$, then there exists a point x_0 in B such that $|Df(x_0)| < 4$.

5.82 It is well known that the derivative of any differentiable function $f : \mathbb{R} \to \mathbb{R}$ has the Darboux property (in particular the image $f'(\mathbb{R})$ is connected). What can be said about the derivative of a function $f : \mathbb{R} \to \mathbb{R}^2$?

5.83 Does there exist a function $f : \mathbb{R}^2 \to \mathbb{R}$ such that
   (i) f is continuous.
   (ii) $\partial f / \partial x$ and $\partial f / \partial y$ exist everywhere.
   (iii) Df does not exist at any point?

5.84 Show that every real convex function defined on a convex subset of $\mathbb{R}^n$ is locally Lipschitz.

5.85 Prove that each bounded convex set V in the plane can be approximated by two-dimensional sections of the cubes $[-1, 1]^N$; that is, for every $\varepsilon > 0$ there exist N and an embedding $f : \mathbb{R}^2 \to \mathbb{R}^N$ such that the gauges (Minkowski functionals) of V and $U = f^{-1}([-1, 1]^N)$ satisfy the inequality $|p_U/p_V - 1| < \varepsilon$.

5.86 Prove the Kronecker theorem: If the real numbers $c_1, \ldots, c_n$ are linearly independent over $\mathbb{Q}$, then the range of the map $k : \mathbb{R} \to \mathbb{T}^n$ defined by $t \mapsto (\exp(ic_1 t), \ldots, \exp(ic_n t))$ is dense in $\mathbb{T}^n$.

   Hence $k(\mathbb{R})$ is a manifold in $\mathbb{T}^n$ whose topology is different from that inherited from $\mathbb{T}^n$.

5.87 Construct a smooth curve $g : \mathbb{R} \to \mathbb{R}^n$ with dense range.

5.88 A differentiable function $f : \mathbb{R}^n \to \mathbb{R}$ has the only critical point $(Df = 0)$ where f attains its local minimum. For which n is this point necessarily the global minimum of f?

5.89 Consider for $a < 0 < b$ the polynomial map $P_{a,b} : \mathbb{R} \times \mathbb{R}^{k-1} \to \mathbb{R}$ given by

$$P_{a,b}(x, y) = \int_0^x s(s - a)(s - b) \, ds + |y|^2$$

Find critical points of $P_{a,b}$. Determine when the set $P_{a,b}^{-1}((-\infty, c])$ is connected.

5.90 Prove the Ekeland variational principle for approximations of the minimization problem:

Let $f: \mathbb{R}^n \to (-\infty, +\infty]$ be a lower semicontinuous function bounded from below and $f \not\equiv +\infty$. For every point x_ε such that $f(x_\varepsilon) \leqslant \inf f + \varepsilon$ and every $c > 0$ there exists y_ε satisfying $f(y_\varepsilon) \leqslant f(x_\varepsilon)$, $|y_\varepsilon - x_\varepsilon| \leqslant c$ and $f(y_\varepsilon) \leqslant f(x) + \varepsilon c^{-1}|x - y_\varepsilon|$ for all $x \in \mathbb{R}^n$.

Deduce from this principle that for every differentiable function $f: \mathbb{R}^n \to \mathbb{R}$ bounded from below and for every $\varepsilon > 0$ there exists y_ε such that $f(y_\varepsilon) \leqslant \inf f + \varepsilon$ and $|Df(y_\varepsilon)| \leqslant \varepsilon^{1/2}$.

5.91 Prove that every closed subset of $\mathbb{R}^n$ is
  (i) the set of zeros of some smooth function $f: \mathbb{R}^n \to \mathbb{R}$
  (ii) the set of critical points of some smooth function $F: \mathbb{R}^n \to \mathbb{R}$

5.92 Construct an example of a function f in $C^1[0,1]$ whose set of critical values $f(\{t: f'(t) = 0\})$ is uncountable.

5.93 Construct a smooth function $f: \mathbb{R}^n \to \mathbb{R}$ such that for every multiindex α, $D^\alpha f$ maps a fixed countable subset $A \subset \mathbb{R}^n$ onto given dense countable set $B_\alpha \subset \mathbb{R}^n$.

5.94 A slightly paradoxical example: construct functions $F, G: \mathbb{R}^2 \to \mathbb{C}$ 2π-periodic in each variable, such that
  (i) $G \in C^4$, $F \in C \backslash C^4$,
  (ii) For some v_1, v_2, $v_1/v_2 \notin \mathbb{Q}$, $f'(t) = g(t)$, where $f(t) = F(v_1 t, v_2 t)$ and $g(t) = G(v_1 t, v_2 t)$.

Note that due to continuity F, G are uniquely determined by f, g.

5.95 A function $f \in C^1(\mathbb{R}^2)$ has the continuous derivatives f_{xx} and f_{yy}. Does it imply the continuity of the mixed derivative f_{xy}?

5.96 A function $f: \mathbb{R}^2 \to \mathbb{R}$ is differentiable separately with respect to each variable and $\partial f/\partial x = \partial f/\partial y$. Does there exist a function g such that $f(x, y) = g(x + y)$?

5.97 Show that the map $F: \mathbb{R}^2 \to \mathbb{R}^2$ defined by the formula $F(x, y) = (x + f(y), y + f(x))$, where f is continuously differentiable and $|DF(t)| \leqslant k < 1$, is a bijection of $\mathbb{R}^2$.

5.98 Let A be a linear transformation of $\mathbb{R}^n$ and $N \in C^2(\mathbb{R} \times \mathbb{R}^n, \mathbb{R}^n)$ satisfy the condition $\lim_{x \to 0} N(\lambda, x)/|x| = 0$ uniformly in λ. Show that if there exists a sequence (λ_k, x_k) such that $Ax_k = \lambda_k x_k + N(\lambda_k, x_k)$, $\lim_{k \to \infty} \lambda_k = \mu$, $\lim_{k \to \infty} x_k = 0$, $x_k \neq 0$, then μ is an eigenvalue of A. Conversely, if μ is a simple eigenvalue of A then there exists a sequence (λ_k, x_k) satisfying the conditions above.

5.99 Does there exist a nonconstant map $F: \mathbb{R}^n \to \mathbb{R}^{n-1}$ such that the limit $\lim_{x \to a} |F(x) - F(a)|/|x - a|$ exists for every $a \in \mathbb{R}^n$?

5.100 Prove that every continuous transformation $F: \mathbb{R}^n \to \mathbb{R}^n$ satisfying the condition

$$\limsup_{|x| \to \infty} |x - F(x)|/|x| < 2$$

is surjective.

5.101 Consider a diffeomorphism F of the unit ball in $\mathbb{R}^n$ such that $F^2 = Id$ and F is the identity in some neighborhood of the origin. Show that $F = Id$. Generalize to the case of a homeomorphism F.

5.102 Let F be a continuous map defined on the unit ball in $\mathbb{R}^n$ such that for every y on the unit sphere $F(y) = my$ for some $m > 1$. Show that F has a fixed point.

5.103 Prove that there is no continuous transformation $F: \{x \in \mathbb{R}^n: |x| \leqslant 1\} \to \{x \in \mathbb{R}^n: |x| = 1\}$ whose restriction to the sphere is the identity.
  Draw as corollaries:
  (i) the Brouwer fixed point theorem.
  (ii) the hairy ball theorem: even-dimensional spheres do not admit non-vanishing tangent C^1 vector fields.

5.104 Show that the system of equations

$$x(x^2 + y^2 + z^2) = f(x, y, z)$$
$$y(x^2 + y^2 + z^2) = g(x, y, z)$$
$$z(x^2 + y^2 + z^2) = h(x, y, z)$$

where f, g, and h are given bounded continuous functions, has a solution in $\mathbb{R}^3$.

5.105 Prove that there is no binary operation $\circ$ in $\mathbb{R}^n$, $n \geqslant 3$, satisfying the following conditions

  (i) $x \circ (cy) = (cx) \circ y = c(x \circ y)$, $c \in \mathbb{R}$

  (ii) $x \circ (y + z) = x \circ y + x \circ z$

  (iii) $x \circ y = 0 \Rightarrow x = 0$ or $y = 0$

  (iv) $x \circ y = y \circ x$

Remark: with the supplementary hypothesis on associativity of $\circ$, the result follows easily from the fundamental theorem of algebra.

5.106 The polynomial $P_0(z) = z^n + a_{01}z^{n-1} + \cdots + a_{0n}$ has simple roots. Prove that in a neighborhood of the point $a_0 = (a_{01}, \ldots, a_{0n})$ the roots of the polynomial $P(z) = z^n + a_1 z^{n-1} + \cdots + a_n$ are smooth (and even analytic) functions of the coefficients $a = (a_1, \ldots, a_n)$. Show that the assumption on simplicity of the roots is essential.

5.107   A system $f = (f_1, \ldots, f_m)$ of smooth functions defined on a domain $G \subseteq \mathbb{R}^n$ is said to be functionally dependent on $S \subseteq G$ if there exists a smooth function g on an open set U containing $f(S)$ such that $g^{-1}(0)$ is nowhere dense and $g(f(x)) = 0$ for $x \in S$. Prove that a system of functions $f = (f_1, \ldots, f_m)$ is functionally dependent on every compact subset of G if and only if the rank of the matrix of derivatives Df is strictly less than m for every x in G.

5.108   Consider the map $F : \mathbb{R}^2 \to \mathbb{R}^3$ defined by $F(x, y) = (x, xy, xe^y)$. Show that there does not exist any function $g \not\equiv 0$ analytic in a neighborhood of the origin in $\mathbb{R}^3$ such that $g \circ F = 0$ in some neighborhood of the origin in $\mathbb{R}^2$.

Remark: The result remains true if we replace e^y by any real function on $\mathbb{R}$ with the entire extension to $\mathbb{C}$ which is not a polynomial.

Corollary: The analytic version of the theorem on functional dependence of the system $(f_1, \ldots, f_m)$ requires (and remains true after) deleting a nowhere dense closed subset of G where g may not be defined.

5.109   Show that a function integrable in the plane with vanishing integrals over each line must vanish a.e.

5.110   Let f be a continuous function on $\mathbb{R}^n$ such that $f(x) = O(|x|^{-n})$ for $|x| \to \infty$. Hence f is summable over each $(n - 1)$-dimensional hyperplane $L \subseteq \mathbb{R}^n$ with respect to the Lebesgue measure m_L. Prove that if $F(L) := \int_L f \, dm_L = 0$ for every hyperplane L, then $f \equiv 0$. Determine f knowing the values $F(L)$ for every L in the case $n = 3$.

5.111   Reconstruct a function $f \in \mathscr{S}(\mathbb{R}^3)$ from its integrals over all lines intersecting the fixed one.

5.112   A continuous function f on $\mathbb{R}^n$ is said to be R-quasiperiodic if the integral of f over each ball of radius R does not depend on the center of this ball. Show that in the one-dimensional case the quasiperiodicity is equivalent to the periodicity. Give examples of nonconstant quasiperiodic functions in several variables.

5.113   Show that the functions

$$u(x) = \int_0^\infty e^{-t} t^{-1/2} \sin tx \, dt, \quad v(x) = \int_0^\infty e^{-t} t^{-1/2} \cos tx \, dt$$

are continuously differentiable and satisfy a system of first order differential equations. Using this fact, compute u and v explicitly.

5.114 Solve the partial differential equation

$$pf + \sum_{j=1}^{n} x_j \frac{\partial f}{\partial x_j} = h$$

on a star-shaped open set in $\mathbb{R}^n$.

5.115 Find all solutions of the equation $u_{xx} + u_{yy} = 2u_{xy}$ which are of the form $u(x, y) = f(x)g(y)$.

5.116 For what functions $f: \mathbb{R} \to \mathbb{R}$ is the function $f \circ u$ harmonic for every harmonic $u: \mathbb{R}^2 \to \mathbb{R}$?

5.117 Find all functions $f \in C^2(\mathbb{R}^+)$ such that $g(x, y, z) = f((x^2 + y^2)/z)$ is harmonic on $\mathbb{R} \times \mathbb{R} \times \mathbb{R}^+$.

5.118 Consider a positive harmonic function u on the unit disk, satisfying $u(0) = 1$. Estimate max $u(1/2)$ and min $u(1/2)$.

5.119 u is a harmonic function on the unit disk and $0 \leqslant u(re^{it}) \leqslant P(r, t)$, where $P(r, t)$ is the Poisson kernel. Show that $u(re^{it}) = cP(r, t)$ for some real c.

5.120 A function is harmonic in the unit disk $D = \{x^2 + y^2 < 1\}$, $u(0, 0) = 1$ and a real function g satisfies $g(0) \leqslant 1/2$. Show that if $g(u) + xu_x + yu_y > 0$ for $(x, y) \in D$, then $u(x, y) > 0$.
Consider special cases $g(t) = 1/2$, $g(t) = 1/2 + t$.

5.121 Let $D = \{z : |z| < 1\}$. The functions $f_1, f_2 : \bar{D} \to \mathbb{R}$ are continuous and superharmonic, subharmonic on D respectively, $f_1(z) \leqslant f_2(z)$ for all $z \in D$. Does there exist a continuous function $h: \bar{D} \to \mathbb{R}$ harmonic on D such that $f_1(z) \leqslant h(z) \leqslant f_2(z)$ on D?

5.122 Prove the Kellogg theorem: Let U be a bounded open set in $\mathbb{R}^n$, $n \geqslant 2$, such that the Dirichlet problem for the Laplacian is solvable in U, and let a continuous real function f on U satisfy the restricted mean value property, that is, for every $a \in U$ there is $r = r(a)$ such that $\{x : |x - a| \leqslant r\} \subseteq U$ and

$$f(a) = \int_{\{x:|x-a|=r\}} f(x) \, dS(x) \Big/ \int_{\{x:|x-a|=r\}} ds(x),$$

where dS is the Lebesgue measure on the $(n-1)$ dimensional sphere.
Prove that f is a harmonic function on U.

5.123 f is a continuous real function on $\mathbb{R}^n$ whose mean

$$M(x, r) = \int_{\{y:|y-x| \leqslant r\}} f(y) \, dy \Big/ \int_{\{y:|y-x| \leqslant r\}} dy, \quad r > 0 \qquad x \in \mathbb{R}^n$$

satisfies the condition

$$\forall x \in \mathbb{R}^n \lim_{r \to 0} \frac{M(x, r) - f(x)}{r^2} = 0$$

Prove that f is a harmonic function. Is the continuity assumption essential?

5.124 Find the equation of the curve in the plane whose tangents are at equal distances from the points $A, -A$.

5.125 Describe the set $\{(u, v): \text{the curves } xu + yv = 1 \text{ and } x^m + y^m = 1 \text{ are tangent}\}$, $m > 1$, $u, v, x, y > 0$.

5.126 Find the envelope of the family of lines $x \cos t + y \sin t = \sin^2 t/\cos t$, $|t| < \pi/2$.

5.127 A particle falls in a vertical plane from rest under the influence of gravity and a force perpendicular and proportional to its velocity. Derive the equation of the trajectory and identify the curve.

5.128 Find the trajectory of the material point on the two-dimensional sphere if the meridians are crossed at equal angles.

5.129 A material point moves inside the paraboloid $x^2 + y^2 = 2pz$, $p > 0$, under the gravity force parallel to the z-axis. Show that:
  (i) This motion is confined to the region bounded by two circles C_1, C_2 described by the equations $z = z_1$ and $z = z_2$.
  (ii) $\omega_1 z_1 = \omega_2 z_2$ and $\omega_1 \omega_2 = g/p$, where ω_1, ω_2 are the angular velocities on the circles C_1, C_2 respectively.

5.130 Let C be a simple closed curve in the plane enclosing a region R which is star-shaped with respect to the point P. Show that C is radially symmetric in P if and only if every line passing through P divides R into subregions of equal areas.

5.131 The curve $C_{p,q}$ is parametrized by $x = (2 - \cos pt) \cos qt$, $y = (2 - \cos pt) \sin qt$, $z = \sin pt$. Show that the curves $C_{p,q}$ and $C_{q,p}$ are isotopic.

5.132 C is a closed convex curve in the plane enclosing a region of area A. Let $w(t)$ denote the width of C in the t-direction. Show that

$$A \leqslant \frac{1}{2} \int_0^{\pi/2} w(t) w(t + \pi/2) \, dt$$

with equality if and only if C is a circle.

5.133 $R(s)$ is a vector equation of a smooth simple closed curve C which bounds some domain of area A, $N(s)$ is the exterior unit normal vector to C.

Define the curve $C(r)$ by $R(s) + rN(s)$. $C(r)$ is (for small r) a curve enclosing a region of area $A(r)$. Show that $dA(r)/dr|_{r=0} = L$, where L is the perimeter of C.

5.134 A function $u \in C^2(\mathbb{R}^n)$ attains its minimal value 0 in an interior point of the domain B. Show that if the second derivatives of u satisfy the inequality $\sum_{i,j=1}^n u_{ij}^2 \leqslant 1$ and if every pair of points in B can be joined by a smooth curve of length less than $\sqrt{2}$ lying in B, then $u < 1$ on B.

5.135 Determine functions $y = y(x) \in C^2$ such that $y(x) > 0$ for $x > 0$, $y' > 0$ and $dY/dX = dy/dx$, where Y and X are the intersection points of the tangent to the curve $y = y(x)$ with the coordinate axes.

5.136 Find the points where the normals to the ellipse $(a \cos t, b \sin t)$ meet producing singularities. Do the same for the parabola.

5.137 The rays emitted from the point at the distance $a = 0.1$ from the center of the unit circle reflect in the circumference. Determine the set of ends of rays of total length $z = 2$.

5.138 Prove that if R is a convex domain in $\mathbb{R}^3$ and $P, Q \in R$ then

$$d(P, Q) = \sup_f \; \inf_{(x,y,z)\in R} \; \frac{|f(P) - f(Q)|}{\sqrt{f_x^2 + f_y^2 + f_z^2}}$$

where the supremum is taken over all differentiable nonconstant functions f.

5.139 Fix on the x-axis in the plane points $x_1 < \cdots < x_n$. Show that if d_j denotes the distance from x_j to the fixed point off the x-axis and $f(x) = (x - x_1) \cdots (x - x_n)$, then

$$\sum_{j=1}^n \frac{d_j^2}{f'(x_j)} = \begin{cases} 1 & \text{if } n = 3 \\ 0 & \text{if } n > 3 \end{cases}$$

5.140 $P(x, y)$, $Q(x, y)$ are the real polynomials and $xP + yQ = 1$ on the unit circle. Show that the system of equations $P = 0$, $Q = 0$ has a solution.

5.141 Let V_n denote the volume of n-dimensional domain defined by the inequality $f(x_1) + \cdots + f(x_n) \leqslant 1$, where f is an even function increasing on $\mathbb{R}^+$ to infinity, $f(0) = 0$. Show that $\lim_{n \to \infty} V_n = 0$.

5.142 The n-dimensional cylinder of maximal volume is inscribed in the n-dimensional ball. Show that the quotient of their volumes tend to $\sqrt{2/\pi e}$.

6

Real Analysis, Measure and Integration

Topics covered include:
lebesgue measure and integral, absolutely continuous functions, moment problems, and ergodic theory.

Note: If not explicitly indicated, the notions of measurability and integrability are understood with respect to the Lebesgue measure m.

6.1 Represent each of the intervals $[0, 1]$, $(0, 1)$, and $(0, 1]$ as the disjoint union of n homeomorphic sets, $n = 1, 2, \ldots, \omega$.

6.2 Given a dense set in the plane does there exist a segment in which this set is dense?

6.3 Describe the structure of the set $\{(x, y) : x, y \in C, \ x - y = d\}$ for the ternary Cantor set C, $-1 \leqslant d \leqslant 1$.

6.4 It is well-known that for the sequence of measurable sets $E_n \subseteq (0, 1)$, $m(E_n) \geqslant c > 0$, there exists a subsequence of indexes (n_k) such that $\bigcap_k E_{n_k} \neq \varnothing$. Prove that one may choose (n_k) such that its upper density in $\mathbb{N}$ is equal to c.

6.5 Show that given two equimeasurable sets $E, F \subseteq [0,1]$ there exists a measure preserving transformation $T: [0,1] \to [0,1]$ injective a.e. such that $T(E) = F$ modulo a set of measure zero.

6.6 Show that each open set in the plane is a disjoint union of closed segments (not reduced to points).

6.7 The family of subsets of measure zero in $[0,1]$ is partially ordered by inclusion. M is a maximal chain in this family. Is the union of M measurable? Can the length of M exceed continuum?

6.8 Define $A_n(x)$ for $x \in (0, 1)$ as the number of "7" among the first n digits of the decimal expansion of x. Prove that the set $\{x: \lim_{n \to \infty} A_n(x)/n \text{ exists}\}$ is measurable but neither open nor closed.

6.9 A is a measurable subset of the interval $[a, b]$, $h > 0$. Show that $(2h)^{-1} \int_a^b m(A \cap (x - h, x + h)) \, dx \leqslant m(A)$.

6.10 Give an example of a measurable set $E \subseteq \mathbb{R}$ such that $E + E$ is not measurable.

6.11 Show that there exists a partition of the interval $(0, 1)$ into countably many measurable sets E_k of second category such that $0 < m(E_k \cap I) < m(I)$ for every nonempty open interval $I \subseteq (0, 1)$.

6.12 Consider a measurable set of reals E such that $m(E \cap I) \geqslant cm(I)$ for every interval $I \subseteq \mathbb{R}$ and some fixed $c > 0$. Show that the complement of E is of measure zero.

6.13 Prove that for every measurable set $E \subseteq [0,1]$, $0 < m(E) < 1$, the relations $\inf_I m(E \cap I)/m(I) = 0$, $\sup_I m(E \cap I)/m(I) = 1$ hold, where sup and inf are taken over all open intervals $I \subseteq [0,1]$.

6.14 Prove the following version of the Steinhaus theorem: If $A \subseteq [0,1]$ is of positive measure and the sequence (d_n) converges to zero, then for almost all $x \in A$, $x + d_n \in A$ for infinitely many n.

6.15 μ is a Borel measure on $[0,1]$. Prove that if for every Borel set $A \subseteq [0,1]$ $m(A) = 1/2$ implies $\mu(A) = 1/2$, then $\mu = m$ on all Borel sets.

6.16 Suppose that $f: \mathbb{R} \to \mathbb{R}$ is locally integrable and $\int_I f = 0$ for every interval of length 1 or $\sqrt{2}$. Show that $f = 0$ a.e.

6.17 Show that for every Borel probability measure μ on $\mathbb{R}$, $m(E) = \int_{\mathbb{R}} \mu(x + E) \, dx$ for each Borel set $E \subseteq \mathbb{R}$.

6.18 Let μ be a Borel probability measure supported on the rationals $\mathbb{Q} \subseteq \mathbb{R}$. Prove that $\chi_{\mathbb{Q}} = \chi_{\mathbb{Q}} * \mu$ on $\mathbb{R}$. Is it possible for other sets as the support of μ?

6.19 μ is a finite Borel measure on $\mathbb{R}$ and $E \subseteq \mathbb{R}$ is a fixed Borel set. Show that there exist two Borel measures μ_1, μ_2 and a sequence of reals (x_n) such that $\mu = \mu_1 + \mu_2$, $\mu_1(x + E) = 0$ for all $x \in \mathbb{R}$ and $\mu_2(\mathbb{R} \setminus \bigcup_n (x_n + E)) = 0$.

6.20 Consider a nonzero Borel measure μ on $\mathbb{R}$ which is quasiinvariant under translations, that is, for all real t the measure μ_t defined by $\mu_t(A) = \mu(t + A)$ is equivalent to μ. Show that μ is equivalent to the Lebesgue measure.

6.21 Consider a translation invariant measure defined on the Borel subsets of $\mathbb{R}$. Is it also invariant under reflections?

6.22 Consider the family A of rectangles in the unit square with their sides parallel to the sides of the square, one of them being of length one. For $P \in A$ define $f(P) = m(P)$. Extend f (to the σ-algebra generated by A) to a measure different than the usual planar Lebesgue measure.

6.23 Consider two positive Borel measures μ_1, μ_2 on the half-plane $\mathbb{R}^2_+ = \{(x, y) : y > 0\}$ such that $\mu_1((x, 0) + C) = \mu_2((x, 0) + C)$ for all $x \in \mathbb{R}$ and the cone $C = \{(x, y) : |x| < cy, \ y > 0\}$. Does it imply the equality $\mu_1(\mathbb{R}^2_+) = \mu_2(\mathbb{R}^2_+)$?

6.24 Construct an example of a set nonmeasurable with respect to any Lebesgue-Stieltjes measure m_g associated to the nondecreasing continuous function g on $\mathbb{R}$: $m_g([a, b)) = g(b) - g(a)$ for $a < b$.

6.25 Consider an example of σ-finite measure on $\mathbb{R}$ which is not regular: let $g(x) = x^{-1/2}$ for $0 < x < 1$ and $g(x) = 0$ elsewhere. Define $f(x) = \sum_{n=1}^{\infty} 2^{-n} g(x + r_n)$, where (r_n) is an enumeration of rationals. Show that $f \in L^1(\mathbb{R})$ and $\mu(E) = \int_E f^2$ defines a σ-finite measure absolutely continuous with respect to the Lebesgue measure although $\mu((a, b)) = \infty$ for all $a < b$.

6.26 For a nonempty subset E of the metric space (X, d) define $n(E, t) = \sup\{|F| : F$ is a finite subset of E such that $d(x, y) > t$ for different $x, y \in F\}$, $n(\varnothing, t) \equiv 0$. Given a positive, strictly decreasing function f on $(0, 1]$ such that $\lim_{t \to 0} f(t) = \infty$ define now $\text{ext}_f(E) = \lim \inf_{t \to 0} n(E, t) / f(t)$, $v_f(E) = \inf\{\sum_{k=1}^{\infty} \text{ext}_f(A_k)\}$, where the infimum is taken over all countable partitions of E into the sets A_k.

Prove that v_f is the metric outer measure, that is, the outer measure satisfying the condition $v_f(A \cup B) = v_f(A) + v_f(B)$ for any pair of separated sets $A, B : d(A, B) > 0$, and that v_f vanishes on countable subsets of X.

Compute v_f for $f(t) = 1/t$, $X = \mathbb{R}$ with the standard metric. Compare v_f with p-dimensional Hausdorff measures.

6.27 Given a bounded open set $G \subseteq \mathbb{R}^n$ show that there does not exist any covering (B_j) of G by open balls in G satisfying

(i) every point in G belongs to infinitely many B_j.

(ii) $\Sigma_{j=1}^{\infty} m(B_j) < \infty$.

Show that there exists a family (B_j) of open balls which covers G such that (i) is satisfied and for every $p > 1$ $\Sigma_{j=1}^{\infty} (m(B_j))^p < \infty$.

6.28 Let X be a metric space without isolated points. Show that if every real valued continuous function on X is uniformly continuous, then X is compact.

6.29 Prove the Baire theorem: Given two real functions g and h on the compact metric space such that g is lower semicontinuous, h is upper semicontinuous, and $g(x) \geq h(x)$ for all x, there exists a continuous function f satisfying the inequalities $h(x) \leq f(x) \leq g(x)$ for all x.

6.30 Does there exist a function $f: \mathbb{R} \to \mathbb{R}$ whose set of discontinuities D is of measure zero but D has an uncountable intersection with every nonempty open set?

6.31 Is it possible for a real function f on $[0, 1]$ to have $f^{-1}(y)$ countably infinite for every y in its image?

6.32 Define the function f on $[0, 1]$ by $f(x) = \Sigma_{r_n < x} 2^{-n}$, where (r_n) is a fixed enumeration of all rationals in $[0, 1]$. Show that f is continuous on $[0, 1] \backslash \mathbb{Q}$ but it is discontinuous on $[0, 1]$.

6.33 The formula $f(x) = \limsup_{n \to \infty} (c_1 + \cdots + c_n)/n$, where $x = 0.c_1 c_2 \ldots$ is the binary expansion of x (with infinitely many zeros), defines the function $f: [0, 1] \to [0, 1]$. Show that f has the Darboux property but it is nowhere continuous.

6.34 Prove that a finite real function f defined on a set A with a complete σ-finite measure is measurable if and only if for every $\varepsilon > 0$ and every subset $C \subseteq A$ of positive measure there is a subset $B \subseteq C$ of positive measure such that the diameter of $f(B)$ is less than ε.

6.35 Prove that each real valued function f on $[0, 1]$ is the composition of two measurable functions g and h.

6.36 Given a continuous real function f on $[0, 1]$ construct a subset $E \subseteq [0, 1]$ of G_δ type such that $f(E) = f([0, 1])$ and f restricted to E is injective.

6.37 Construct a subset $E \subseteq (0, 1)$ of measure zero such that for every Riemann integrable function on $(0, 1)$ some of its continuity points belong to E.

6.38 A measurable real function f satisfies the equality $f(qx) = qf(x)$ for all integer q and real x. Show that $f(x) = bx$ a.e. in $\mathbb{R}$ for some real b.

Caution: "a.e." is important here.

Apply this result to prove that each measurable additive function on $\mathbb{R}$ is linear.

6.39 For a measurable additive function $f: \mathbb{R} \to \mathbb{R}$ consider the functions $g(t) = f(t) - tf(x)/x$ for fixed $x \neq 0$ and $h(t) = (1 + |g(t)|)^{-1}$. Show that g, h are x-periodic and

$$\int_0^x h(t)\, dt = \int_0^x (1 + 2|g(t)|)^{-1}\, dt$$

hence

$$\int_0^x |g(t)|(1 + 2|g(t)|)^{-1} h(t)\, dt = 0$$

Of course, this implies that f is linear.

6.40 It is well-known that each additive discontinuous function must be unbounded on every open interval. Does it take all real values on every open interval?

6.41 Consider an additive map $T: \mathbb{R}^n \to \mathbb{R}^m$ such that $T^{-1}(U)$ is measurable for every open set $U \subseteq \mathbb{R}^m$. Prove that T is continuous, hence linear.

6.42 Show that a measurable function f on $\mathbb{R}$ satisfying the inequality $|f(x + y)| \leq G(f(x), f(y))$ for some continuous function of two variables G is locally bounded.

6.43 Prove the C.T. Ionescu Tulčea theorem: If the real function f satisfies the inequality $f(x + y) \leq g(x) + h(y)$ for arbitrary real x and y, where g and h take the value $+\infty$ on the sets of measure zero only, and at least one of g and h is measurable, then f is locally bounded from above.

6.44 Prove that a measurable function g on $\mathbb{R}$ having arbitrarily small periods must be constant a.e.

6.45 Is it possible to construct two nonconstant periodic functions of incommensurable periods such that their sum is periodic?

6.46 Does there exist a periodic measurable function $f: \mathbb{R} \to \mathbb{R}$ such that $f + g$ is periodic for every continuous function $g: \mathbb{R} \to \mathbb{R}$? Answer the same question without the measurability assumption on f.

6.47 Define the function f by the formula $f(x) = \sum_{k=1}^{\infty} k^{-2}|x - r_k|^{-1/2}$, where (r_k) is an enumeration of all rationals in $[0, 1]$. Show that f is finite a.e. and locally integrable.

6.48 Assume that $\sum_{n=1}^{\infty} a_n^{1-c} < \infty$ for positive a_n and some $c > 0$. Prove that if $(x_n) \subseteq [0, 1]$, then the series $\sum_{n=1}^{\infty} a_n/(x - x_n)$ converges absolutely a.e. in $[0, 1]$. Show that for $c = 0$ this conclusion does not hold.

6.49 (r_n) is an enumeration of the rationals in $[0, 1]$ and $r_n = p_n/q_n$ is the representation of this fraction with relatively prime positive p_n, q_n. Study the convergence of the functional sequence $f_n(x) = \exp(-(p_n - q_n x)^2)$ on $[0, 1]$.

6.50 Given a dense countable set $D = (r_m)$ in $\mathbb{R}$ and a positive sequence (a_n) converging to zero construct continuous functions $f_n: \mathbb{R} \to \mathbb{R}^+$ such that $\lim_{n \to \infty} f_n(x) = 1/a_m$ if $x = r_m \in D$ and $= \infty$ if $x \notin D$. Do such functions exist if (a_n) does not converge to zero?

6.51 The function $f: [a, b] \to \mathbb{R}$ is called quasicontinuous if there exists a sequence (K_n) of compacts in $[a, b]$ and a sequence of continuous functions (f_n) such that $f_{|K_n} = f_{n|K_n}$ and $\lim_{n \to \infty} m(K_n) = b - a$.
Show that the quasicontinuity is equivalent to measurability. Prove that for a bounded quasicontinuous function f the limit of the Riemann integrals $\int_a^b f_n$ exists and equals the Lebesgue integral $\int_a^b f \, dm$.

6.52 Prove the Osgood theorem (without use of the Baire category theorem): If a sequence (f_n) of continuous functions on $(0, 1)$ is pointwise bounded, then (f_n) is uniformly bounded on a nonempty interval.

6.53 Prove in an elementary way (without the Lebesgue theory) the Arzelà theorem: If Riemann integrable functions f_n on $[0, 1]$ are uniformly bounded and f_n converge to f pointwise, then $\lim_{n \to \infty} \int_0^1 f_n = \int_0^1 f$.

6.54 Construct an example of a pointwise converging sequence of measurable functions on $(0, 1)$ such that for every subset $E \subseteq (0, 1)$ of full measure this sequence does not converge uniformly on E.
This example would show that the conclusion of the Egorov theorem cannot be improved substantially.

6.55 Let (P_n) be a sequence of polynomials of degrees k_n and P_n converging uniformly in $[0, 1]$ to a function which is not a polynomial. Show that $\lim_{n \to \infty} k_n = \infty$ and $\lim_{n \to \infty} m\{x : P_n(x) = f(x)\} = 0$.

6.56 Give an example of a bounded measurable function f on $[0, 1]$ such that the sequence $f_n(x) = f(x - 1/n)$ diverges a.e.

6.57 Let f and g be nonnegative measurable functions on the finite measure spaces (X, μ), (Y, ν) respectively and $\int_X f \, d\mu = \int_Y g \, d\nu < \infty$. The condition $C_{f,g}$ reads: for every measurable subset $E \subseteq Y$ $\int_X \min(f, \nu(E)) \geq \int_E g$. The condition $C_{g,f}$ is obtained exchanging the roles of f, g and X, Y. Show that $C_{f,g}$ and $C_{g,f}$ are equivalent.

6.58 Let P denote the partition of $[0, 1]$ into a finite number of measurable sets E_j, $|P| = \max m(E_j)$ and let x_j belongs to E_j. Characterize the functions f for which the limit $\lim_{|P| \to 0} \Sigma_j f(x_j) m(E_j)$ exists.

6.59 Define for a continuous function f on $\mathbb{R}^+$ $h(x) = \sup_{s \geqslant 0} \int_x^{x+s} f(t)\,dt$. Show that $h(x)$ converges to zero as x tends to infinity if and only if there is a Riemann integrable function $g \geqslant 0$ such that $f(x) \leqslant g(x)$ for all x and the integral $\int_0^\infty g(t)\,dt$ is finite.

6.60 A nonnegative measurable function f on $\mathbb{R}$ satisfies for all $a < b, R > 0$ the condition $m((a, b) \cap \{x : f(x) \geqslant R\}) > 0$. Is $\int_{-\infty}^\infty f(x)\,dx = \infty$ necessarily true?

6.61 Is it possible for a measurable function $f : \mathbb{R}^+ \to \mathbb{R}$ that $|f(x)| \geqslant \int_0^x |f|^p$ holds for all $x \geqslant x_0$ and some $p > 1$?

6.62 The functions f and g belong to $L^\infty(\mathbb{R}^+)$ and satisfy the condition $\int_0^\infty (|f(x)| + |g(x)|)x^{-1}\,dx < \infty$. Show that $\int_0^\infty |f(xy)g(1/y)|y^{-1}\,dy < \infty$ a.e.

6.63 The function f is measurable and bounded on $\mathbb{R}^+$. Consider $F(x) = (f * f)(x) = \int_0^x f(t)f(x - t)\,dt$. Is F positive in some neighborhood of the origin?

6.64 Define for an integrable positive function f_0 on $[0, 1]$ $f_{n+1}(x) = (\int_0^x f_n(t)\,dt)^{1/2}$, $n = 0, 1, 2, \ldots$. Assuming that $f_0(x) \geqslant f_1(x)$ find $\lim_{n \to \infty} f_n(x)$. Is the hypothesis $f_0(x) \geqslant f_1(x)$ important?

6.65 Find a positive locally bounded function f on $\mathbb{R}^+$ which is upper semicontinuous and $\int_0^\infty f = \infty$ though $\sum_{n=1}^\infty f(nx) < \infty$ for every $x > 0$.

6.66 Let $\mathscr{F}_\varepsilon$, $0 < \varepsilon < 1$, denote the family of all nonnegative integrable functions on $\mathbb{R}^+$ such that $f(x)f(y) \leqslant \varepsilon^{|x-y|}$. Compute $s_\varepsilon = \sup_{f \in \mathscr{F}_\varepsilon} \int_0^\infty f(x)\,dx$.

6.67 Let $F, G \in L^1(\mathbb{R})$ and $\int F = \int G = 1$. Show that for every $0 < t < 1$ there exists a measurable set E satisfying $\int_E F = \int_E G = t$.

6.68 Consider integrable functions $f_1, \ldots, f_n$ on $(0, 1)$, $\int_0^1 f_j = 0, j = 1, \ldots, n$. Show that for every $0 < a < 1$ there exists a subset $E \subseteq (0, 1)$ such that $m(E) = a$ and $\int_E f_j = 0$ for all j.

6.69 Consider the set of Riemann (or Lebesgue) integrable functions on $(0, 1)$ such that $0 \leqslant f(x) \leqslant 1$. Determine the extreme points of this set and all functions representable as weak limits of these extreme points (f is a weak limit of (f_n) if for every integrable function h $\int_0^1 fh = \lim_{n \to \infty} \int_0^1 f_n h$).

6.70 Suppose that E is a measurable subset of $\mathbb{R}$, (f_n) a sequence of real integrable functions such that $\int_F f_n$ converges to zero for every measurable subset F of E. Does it imply that $\int_E |f_n|$ converge to zero?

6.71 Define for an integrable real measurable function f with support in $[0, 1]$ $Tf(y) = \int_\mathbb{R} (f(y - x) + f(1 + y - x))f(x)\,dx$ for $y \in [0, 1]$ and $Tf(y) = 0$ elsewhere. Show that $f \equiv 0$ and $f \equiv 1$ are the only solutions in

$L^2(0, 1)$ of the equation $Tf = f$. Do there exist solutions which are not square integrable?

6.72 Show that for each real function $f \in L^2(0, 1)$

$$\lim_{n \to \infty} n \sum_{k=0}^{n-1} \left(\int_{k/n}^{(k+1)/n} f \right)^2 = \int_0^1 f^2$$

6.73 Show that $f \in L^2(0, 1)$ if and only if $f \in L^1(0, 1)$ and there exists a nondecreasing function h such that for all $r < s$ $|\int_r^s f(x) \, dx|^2 \leqslant (h(s) - h(r))(s - r)$.

6.74 Let the function $F : (0, 1] \to \mathbb{R}^+$ satisfy the estimate $F(x) \leqslant (f(x) - f(y))/(x - y)$ for an increasing function f, $f(0) = 0$, $f(1) = 1$ and all $0 < y < x \leqslant 1$. Prove that $h(t) = \sup_{t \leqslant x \leqslant 1} \inf_{0 < y < x} (f(x) - f(y))/(x - y)$ is the least decreasing integrable majorant of F. Moreover $\int_0^1 h(t) \, dt = 1$.

6.75 Consider a 1-periodic function f integrable over $[0, 1]$. Prove that

$$\lim_{k \to \infty} 2^{-k} \sum_{j=1}^{2^k} f(x + j/2^k) = \int_0^1 f(t) \, dt$$

for almost all x. Observe however that the sequence $k^{-1} \sum_{j=1}^k f(x + j/k)$ may diverge everywhere.

6.76 Consider the set $\{g \in L^1(0, 1) : |g| \leqslant h \text{ a.e.}\}$ for a positive measurable function h on $(0, 1)$. Show that this set is closed in $L^1(0, 1)$. Moreover $h \in L^1(0, 1)$ if this set is compact. The converse is not true: $h \in L^1(0, 1)$ implies only the weak compactness of the considered set.

6.77 The family of functions $\mathscr{F} \subseteq L^1(\mu)$ is uniformly integrable, that is, for every $\varepsilon > 0$ there is $k = k(\varepsilon)$ such that $\int_{\{|f| > k\}} |f| \, d\mu < \varepsilon$ for all $f \in \mathscr{F}$. Does there exist an integrable function h such that $|f| \leqslant h$ a.e. for all $f \in \mathscr{F}$?

6.78 Prove that for every function f integrable over $\mathbb{R}$ there exists an increasing function F on $\mathbb{R}^+$ such that $\lim_{t \to \infty} F(t) = \infty$ and $\int |f(x)| F(|f(x)|) \, dm(x) < \infty$. Show that F can be chosen the same for a family of functions $\mathscr{F}$ if and only if the functions in $\mathscr{F}$ have uniformly absolutely continuous integrals.

6.79 Consider a sequence of functions (f_n) bounded in $L^r(0, 1)$, $0 < r \leqslant \infty$, converging to f a.e. Prove that for each $0 < p < r$, f_n converge to f in $L^p(0, 1)$.

6.80 Let μ be a nonatomic σ-finite measure on the space X. Define $\rho(f) = \sup_E \{\int_E |f| \, d\mu : \mu(E) = 1\}$, $L_\rho = \{f : \rho(f) < \infty\}$. Show that
  (i) L_ρ with the norm ρ is a Banach space.
  (ii) $f \in L_\rho$ if and only if $f = f_1 + f_2$ for some $f_1 \in L^1$, $f_2 \in L^\infty$.

(iii) $\rho(f) = \inf\{\|f_1\|_1 + \|f_2\|_\infty : f = f_1 + f_2, f_1 \in L^1, f_2 \in L^\infty\}$.

(iv) $L^p \subseteq L_\rho$ for all $1 \leqslant p \leqslant \infty$.

6.81 Define for a function $f:[a, b] \to \mathbb{R}$ and partition $P = \{a = u_0 < u_1 < \cdots < u_n = b\}$

$$V_1(f, [a, b], P) = \sum_{j=1}^{n-1} \operatorname{osc}(f, [u_j, u_{j+1}])$$

Show that $\sup_P V_1(f, [a, b], P) = V(f, [a, b])$.

Prove that f is equal a.e. to a function of bounded variation if and only if $\int_a^b |f(x + h) - f(x)| \, dx = O(h)$ for $h \to 0$.

6.82 Using the identity $\int_X (f(x) - \int_X f(y) \, d\mu(y))^2 \, d\mu(x) = \int_X f(x)^2 \, d\mu(x) - (\int_X f(x) \, d\mu(x))^2$ valid for every $f \in L^2(X, \mu)$, $\mu(X) = 1$, show that for $0 \leqslant x \leqslant 1$ and positive integer n $\sum_{k=1}^n (x - k/n)^2 \binom{n}{k} x^k (1 - x)^{n-k} \leqslant 1/4n$.

6.83 Prove that the length of the graph of a nondecreasing continuous function f on $[0, 1]$, $f(0) = 0$, $f(1) = 1$, is less or equal to 2 with the equality if and only if $f'(x) = 0$ a.e.

6.84 Define for a continuous function $f : [a, b] \to \mathbb{R}$ the Banach indicatrix of f on the interval $[m, M] = [\min f, \max f]$ by $N(y) =$ number of solutions of the equation $f(x) = y$. Prove that $\int_m^M N(y) \, dy = V(f, [a, b])$.

6.85 Give an example of a real function on $[0, 1]$ with a bounded derivative which is not Riemann integrable.

6.86 Construct a real function whose derivative exists and is finite everywhere but it is not Lebesgue integrable.

6.87 Show that the Riemann function defined by $f(x) = 0$ for irrational x and $f(p/q) = 1/q$ if p, q are relatively prime, $q > 0$, is nowhere differentiable.

6.88 Let (q_n) be an enumeration of rationals in $[0, 1]$ and $x \geqslant 2$. Study continuity and differentiability properties of the function $f(x) = m(\bigcup_{n=1}^\infty (q_n - x^{-n}, q_n + x^{-n}))$.

6.89 Define for $0 < r \neq 1$ the function $f:[0, 1] \to \mathbb{R}$ as follows: If $x = \sum_{n=0}^\infty 2^{-a_n}$, where (a_n) is a strictly increasing sequence of positive integers, then $f(x) = \sum_{n=0}^\infty r^n (1 + r)^{-a_n}$, $f(0) = 0$.

Show that f is continuous and strictly increasing and its derivative is zero a.e.

6.90 Find a real function which is not monotone on any open interval but has a bounded derivative everywhere.

6.91 Let $g:[0, 1] \to [0, 1]$ be the Cantor-Lebesgue function. Define the measure μ on the Cantor set C by $\mu(S) = m(g(S))$ for $S \subseteq C$. It is well-known that $g'(x) = 0$ m-a.e. in $[0,1]$. Prove that $g'(x) = \infty$ μ-a.e. in C.

6.92 Construct a function $f:[0,1] \to \mathbb{R}$ such that its derivative exists everywhere and it is discontinuous for all rationals.

6.93 Does the inequality $x^{-1} \int_0^x f(t)\,dt \geq f(x)$ a.e. in $\mathbb{R}^+$ imply that f decreases?

6.94 Prove that if the continuous function $f:\mathbb{R} \to \mathbb{R}$ satisfies the condition $\lim \sup_{h \to 0} (f(x+h) - f(x))/h \geq 0$ a.e., then f is increasing.

6.95 Show that an everywhere differentiable function f strictly increases if and only if $f' \geq 0$ and the set $\{x : f'(x) > 0\}$ has empty interior.

6.96 Prove that if the right derivative of a continuous function $f:[0,1] \to \mathbb{R}$ vanishes everywhere, then f is constant.

Verify that if $\int_0^1 |f(x+h) - f(x)|\,dx = o(h)$ if h tends to zero, then f is constant a.e.

6.97 Define the function f on $[0,1]$ as follows: If $0.a_1a_2a_3 \ldots$ is the binary expansion of t (containing infinitely many zeros), then $f(t) = \sum_{n=1}^{\infty} a_n/4^n$. Prove that f is continuous on the right, its right derivative is zero, but f is not constant on any open interval.

6.98 Suppose that $f:\mathbb{R} \to \mathbb{R}$ is differentiable everywhere and
  (i) $f'(x) = 0$ for all $x \notin \mathbb{Q}$.
  (ii) $f'(x) = 0$ for all $x \in \mathbb{Q}$.
Does it imply that f is constant?

6.99 Prove that $f:[a,b] \to \mathbb{R}$ satisfies the Lipschitz condition if and only if for every $\varepsilon > 0$ there exists a $\delta > 0$ such that for arbitrary (even overlapping) subintervals $[a_k, b_k] \subseteq [a,b]$, $k = 1,2,\ldots,n$, satisfying $\sum_{k=1}^{n}(b_k - a_k) < \delta$ the inequality $\sum_{k=1}^{n} |f(b_k) - f(a_k)| < \varepsilon$ holds.

Compare this property with the definition of absolute continuity.

6.100 Show that a function $f:[a,b] \to \mathbb{R}$ is absolutely continuous if and only if there exists a sequence of Lipschitz functions (f_n) such that $V(f - f_n, [a,b])$ converges to zero.

6.101 Prove that if a continuous function $f:[a,b] \to \mathbb{R}$ has a derivative which is finite except for a countable subset in (a,b) and $f' \in L^1(a,b)$, then f is absolutely continuous.

6.102 Show that a function f is absolutely continuous on a finite interval if and only if it is continuous, has bounded variation, and $f(E)$ is of measure zero for every E of measure zero.

Give examples showing that each of these conditions is important.

6.103 Construct a continuous function differentiable at given points $a_1, \ldots, a_n \in \mathbb{R}$ only.

6.104 Show that the Van der Waerden function $\sum_{n=0}^{\infty} [\![2^n x]\!] / 2^n$, where $[\![x]\!]$ denotes the distance from x to the integers, does not have derivative (even one-sided) at any point.

Prove also that the function $\sum_{n=0}^{\infty} [\![2^{3^n} x]\!] / 2^n$ has an infinite one-sided derivative at each point.

6.105 Define the generalized derivative of the function f as $\lim_{h \to 0} 2(h^{-1} \int_x^{x+h} f(t)\, dt - f(x))/h$. Show that there exist functions nowhere differentiable in this sense.

6.106 Let $0 < b < 1$ and a be an odd integer such that $ab > 1 + 3\pi/2$.

Prove that the Riemann function $f(x) = \sum_{k=0}^{\infty} b^k \cos(a^k \pi x)$ is continuous and bounded on $\mathbb{R}$ but it does not have finite derivative at any point.

Remark: the same holds even if $ab \geqslant 1$. The differentiability question for the Hardy function $\sum_{k=1}^{\infty} \sin k^2 x / k^2$ is far more delicate, see S. Itatsu, *Proc. Japan Acad. 57A*, 1981, 492–495 and M. Hata, *J. Analyse Math. 51*, 1988, 62–90.

6.107 Verify that the Weierstrass curve $x(t) = \sin t$, $y(t) = \sum_{n=1}^{\infty} 2^{-n} \cos(3^n t)$ has no tangent at any point.

6.108 Define the piecewise linear function f in the interval $[0, 1]$ so that $f([0, 1/3]) = \{0\}$, $f([2/3, 1]) = \{1\}$, $f'((1/3, 2/3)) = \{3\}$ and extend f to the even function of period 2. Show that the curve $x(t) = \sum_{n=0}^{\infty} 2^{-n-1} f(3^{2n} t)$, $y(t) = \sum_{n=0}^{\infty} 2^{-n-1} f(3^{2n+1} t)$ is a Peano curve.

6.109 Construct a continuous function on $[0, 1]$ with dense set of local maxima.

Prove that the set of such functions is Borel in $C[0, 1]$ with its complement of first category.

6.110 Given a real continuous function f on $\mathbb{R}$ does there exist a strictly increasing function $g: \mathbb{R} \to \mathbb{R}$ such that $g \circ f$ is differentiable everywhere?

6.111 Prove the following simplified version of the Denjoy-Saks-Young theorem: Define for arbitrary function $f: \mathbb{R} \to \mathbb{R}$

$$D_{\pm} f(x) = \lim_{\substack{\sup \\ \inf \\ 0 < h+k \to 0 \\ h \geqslant 0, k \geqslant 0}} \frac{f(x+h) - f(x-k)}{h+k}$$

There exists a set of measure zero such that on its complement either the derivative numbers of f $D_{\pm} f$ are $\pm\infty$ or the derivative of f exists and it is finite.

6.112 Suppose that g is a measurable bounded function defined on $S \subseteq \mathbb{R}^n$, $m(S) < \infty$, $a = \inf g$, $b = \sup g$, and h is absolutely continuous on $[a, b]$.

Prove the following formula

$$\int_S h \circ g = m(S)h(b) - \int_a^b h'(t)m(g^{-1}([a,t]))\,dt$$

6.113 (f_n) is a sequence of increasing functions on $[0,1]$ such that $\lim_{n\to\infty} f_n(x) = 1$ for all $x \in [0,1]$. Show that $\liminf_{n\to\infty} f_n'(x) = 0$ a.e. Give an example where $\lim_{n\to\infty} f_n'(x) = \infty$ for some x.

6.114 Prove the Fubini theorem for series: If (f_n) is a sequence of nondecreasing functions defined on an interval such that $\Sigma_{n=1}^\infty f_n(x) = s(x)$ is finite, then $s'(x) = \Sigma_{n=1}^\infty f_n'(x)$ a.e.

6.115 Define for a function $f:[a,b] \to \mathbb{R}$ of bounded variation $V_f(x) = V(f,[a,x])$. Show that $V_f'(x) = |f'(x)|$ a.e. However there might exist points where V_f has the derivative and f does not have it and vice versa.

6.116 Given a sequence (a_n), $n \geqslant 0$, define the differences $\Delta^0 a_n = a_n$, $\Delta^{k+1} a_n = \Delta^k a_n - \Delta^k a_{n+1}, k = 0, 1, 2, \dots$. Prove that a necessary and sufficient condition for the existence of an increasing function g such that $\int_0^1 x^n \, dg(x) = a_n$, $n = 0, 1, 2, \dots$, is $\Delta^k a_n \geqslant 0$ for all k, n.

6.117 Prove that there exists a function g of bounded variation such that $\int_0^1 x^n \, dg(x) = a_n$ if and only if $\Sigma_{k=0}^n \binom{n}{k} |\Delta^{n-k} a_k| \leqslant C$ for all n and some C.

6.118 Assume that for a real sequence (t_n) $\Sigma_{n=0}^N a_n t_n \geqslant 0$ whenever $\Sigma_{n=0}^N a_n e^{nx} \geqslant 0$ for all x. Prove that there exists the unique Borel measure μ on $[0,1]$ with $t_n = \int_0^1 e^{nx} \, d\mu(x)$.

6.119 Show that if $f \in L^\infty(0,1)$ and $\int_0^1 t^n f(t) \, dt = 0$ for all positive integers n, then $f = 0$ a.e.

6.120 Does there exist a complex Borel measure μ on $[-1,1]$ such that $i^n = \int_{-1}^1 t^n \, d\mu(t)$ for all positive integers n?

6.121 Let (a_n), $n \geqslant 0$, be a complex sequence satisfying for all complex sequences (b_n), all positive integer N and some $1 < p < \infty$ $|\Sigma_{n=0}^N a_n b_n|^p \leqslant \int_0^1 |\Sigma_{n=0}^N b_n t^n|^p \, dt$. Prove that there exists the unique function $f \in L^q(0,1)$, $1/p + 1/q = 1$, such that $a_n = \int_0^1 t^n f(t) \, dt$.

6.122 The Stieltjes moment problem: Prove that a sequence (a_n) of positive numbers is the sequence of moments of a measure on $\mathbb{R}^+$ if and only if $\Sigma_{n,m=0}^N b_m \bar{b}_n a_{n+m} \geqslant 0$ and $\Sigma_{n,m=0}^N b_m \bar{b}_n a_{n+m+1} \geqslant 0$ for all positive integers N and all $(b_1, \dots, b_N) \in \mathbb{C}^N$.

Moreover if $a_n \leqslant CD^n(2n)!$ for some constants C and D, then the solution of this problem is unique.

6.123 A measurable function $f: \mathbb{R} \to \mathbb{R}$ satisfies for every positive integer n the inequality $\int_{\mathbb{R}} |x|^n |f(x)| \, dx \leq 1$. Does it imply that $f = 0$ a.e. on $\mathbb{R} \setminus [-1, 1]$?

6.124 Consider a finite measure space (X, μ) and an integrable function $f: X \to \mathbb{R}^+$. Show that if $T: X \to X$ is a measure preserving bijective transformation, then $\int_X (f/f \circ T) \, d\mu \geq \mu(X)$.

6.125 μ is a probability measure on the space X such that for every $0 < t < 1$ there exists a subset $Y \subseteq X$ with $\mu(Y) = t$. The transformation $T: X \to X$ is measure preserving and $T^k = I$ for some integer $k > 0$. Prove that T is not ergodic.

6.126 Verify that the successive positive integers $a_1, a_2, a_3, \ldots$ in the continuous fraction of $x \in (0, 1)$

$$x = \cfrac{1}{a_1 + \cfrac{1}{a_2 + \cfrac{1}{a_3 + \cdots}}}$$

are determined by the formulas $a_1 = a_1(x) = [1/x]$, $a_2 = a_1(Tx)$, $a_3 = a_1(T^2 x), \ldots$, where $Tx = 1/x - [1/x]$.

Show that T is measure preserving and ergodic on $((0, 1)$, $((1 + x) \log 2)^{-1} \, dx)$. Apply the ergodic theorem to conclude that the frequency of the appearance of the integer k in (a_n) is equal to $\log((k + 1)^2/k(k + 2))/\log 2$.

6.127 Define the transformation $T: [0, 1] \to [0, 1]$ as follows: if $x = 0.a_1 a_2 a_3 \ldots$ is the decimal expansion, then $Tx = 0.a_{n+1} a_{n+2} \ldots$ if a_n is the first "7" in this expansion and $Tx = x$ if it does not contain any "7." Show that T is strongly mixing.

6.128 Prove that if T is a measure preserving ergodic homeomorphism of the unit square S, then there exists a subset $D \subseteq S$ such that $S \setminus D$ is of first category and measure zero, and for every $p \in D$ the sequence of iterates p, $T(p)$, $T^2(p), \ldots$ is dense in S.

Compare with the Poincaré recurrence theorem (see V. I. Arnold, *Mathematical Methods of Classical Mechanics*, Springer, Berlin, 1978 or I. P. Cornfeld, S. V. Fomin, Ya. G. Sinai, *Ergodic Theory*, Springer, Berlin, 1982).

6.129 T is a bijective transformation of the probability measure space (X, μ), T, T^{-1} are measure preserving and T does not have invariant sets of measure different than 0, 1. Prove that the first return function, defined for the set of positive measure A by $n(x) = \min\{k : T^k(x) \in A,\ k \geq 1\}$ and $n(x) = 0$ if $T^k(x) \notin A$ for $k \geq 1$ is measurable and $\int_A n(x)\,d\mu(x) = 1$ (the Kac formula).

6.130 Show that the uncountable product of the intervals $(0,1]$ is not the Borel set in the product of the same cardinality of the intervals $[0,1]$.

6.131 Construct a set of cardinality continuum of continuous regular probability measures on $[0,1]$ which are mutually singular.

7

Analytic Functions

Topics covered include:
elementary properties, estimates, zeros, entire functions, singular points, continuation, varia, and several complex variables.

7.1 Determine the set of cluster points of the set $\{(1^{ia} + 2^{ia} + \cdots n^{ia})/n : n \in \mathbb{N}\}$, where $a \in \mathbb{R}$.

7.2 Prove that if the series $\sum_{n=0}^{\infty} a_n \exp(c_n z)$, $a_n \in \mathbb{C}$, $c_n \in \mathbb{R}$, absolutely converges for $z_1, \ldots, z_m \in \mathbb{C}$, then it absolutely converges in the smallest closed convex polygon containing $z_1, \ldots, z_m$.

7.3 Determine the radius of convergence of the series $\sum_{n=0}^{\infty} a_n z^n$, where $a_n = k!$ if $n = k^2$, $a_n = 0$ otherwise.

7.4. Study the convergence of the series

$$1 + \frac{z}{2!} + \frac{z(z+1)}{3!} + \cdots + \frac{z(z+1)\cdots(z+n)}{(n+2)!} + \cdots$$

7.5 Find the sum of the series $\sum_{n=0}^{\infty} s_n z^n$, where $s_n = \sum_{k=0}^{n} k\binom{n}{k}$.

7.6 Where does the series $\sum_{n=0}^{\infty} \sqrt{n} \exp(n(i^n z - 1))$, $z \in \mathbb{C}$, converge absolutely?

7.7 Study the pointwise and absolute convergence of the series $\sum_{n=1}^{\infty} n^{-1} \sin nz$ and $\sum_{n=1}^{\infty} e^{-n} \sin nz$.

7.8 Show that the sequence (a_n) with general term

$$1 - \frac{n-1}{1!} + \frac{(n-2)^2}{2!} - \frac{(n-3)^3}{3!} + \cdots + (-1)^{n-1} \frac{1}{(n-1)!}$$

converges to zero.

7.9 Compute the limit $\lim_{x \to 1-0} (1-x) \sum_{n=0}^{\infty} a_n x^n$, $x \in \mathbb{R}$, $a_n \in \mathbb{C}$, if for some N and all k, $a_{kN}, \ldots, a_{(k+1)N-1}$ are a permutation of $a_0, \ldots, a_{N-1}$.

7.10 Prove the Appell power series comparison theorem: If $p_n > 0$, $\sum_{n=0}^{\infty} p_n = \infty$ and the radius of convergence of the power series $\sum_{n=0}^{\infty} p_n z^n$ is equal to 1, $\lim_{n \to \infty} a_n/p_n = s$, then the radius of convergence of the series $\sum_{n=0}^{\infty} a_n z^n$ is also equal to 1. Moreover

$$\lim_{r \to 1} \sum_{n=0}^{\infty} a_n r^n \bigg/ \sum_{n=0}^{\infty} p_n r^n = s$$

Prove the same result supposing, instead of $\lim_{n \to \infty} a_n/p_n = s$, that $\lim_{n \to \infty} (a_0 + \cdots + a_n)/(p_0 + \cdots + p_n) = s$ only.

7.11 Show that the Appell theorem implies:
  (i) The Abel lemma: if $\sum_{n=0}^{\infty} a_n = s$, then $\lim_{r \to 1} \sum_{n=0}^{\infty} a_n r^n = s$.
  (ii) The Frobenius theorem: if $s_n = a_0 + \cdots + a_n$ and $\lim_{n \to \infty} (s_0 + \cdots + s_n)/n = s$, then $\lim_{r \to 1} \sum_{n=0}^{\infty} a_n r^n = s$.

7.12 Prove the Fejér theorem: If the power series $f(z) = \sum_{n=0}^{\infty} a_n z^n$ converges for $|z| < 1$, $\lim_{r \to 1} f(r) = s$ and $\sum_{n=0}^{\infty} n|a_n|^2 < \infty$, then $\sum_{n=0}^{\infty} a_n = s$.

7.13 (a_n) is a complex sequence such that $\limsup_{n \to \infty} |a_{n+1}/a_n| = A$. What can be said about the radius of convergence of the power series $\sum_{n=0}^{\infty} a_n z^n$? Answer the same question assuming that $\lim_{n \to \infty} |a_{n+1}/a_n| = A$.

7.14 The analytic function $f(z) = \sum_{n=0}^{\infty} a_n z^n$ has on the boundary of its disk of convergence the only singular point z_0 which is a pole. Show that $\lim_{n \to \infty} a_n/a_{n+1} = z_0$.

7.15 The radius of convergence of the power series $f(z) = \sum_{n=0}^{\infty} a_n z^n$ is equal to R and the complex sequence (b_n) satisfies the condition $\lim_{n \to \infty} b_n/b_{n+1} = q$, $|q| < R$. Show that if $c_n = \sum_{k=0}^{n} a_k b_{n-k}$, then $\lim_{n \to \infty} c_n/b_n = f(q)$.

7.16 The series $f(z) = \sum_{n=0}^{\infty} a_n z^n$ converges for $|z| < R$ and θ is irrational. Define $A_n(z) = \sum_{k=0}^{n} f(e^{2\pi i k \theta} z)/(n+1)$. Show that $\lim_{n \to \infty} A_n(z) = f(0)$.

7.17   The series $1 + \Sigma_{n=1}^{\infty} a(a + n)^{n-1} x^n e^{-nx}/n!$, $a \neq 0$, converges uniformly for all $x \geqslant 0$. It represents the function e^{ax} for $0 \leqslant x \leqslant 1$ and a different analytic function for $x > 1$. Explain this slightly astonishing situation.

7.18   Let $f(z) = z(z + 1)$, $S_1(z) = z$, $S_{n+1}(z) = f(S_n(z))$. Find the domain of convergence of the series $\Sigma_{n=1}^{\infty} (S_n(z) + 1)^{-1}$ and its sum.

7.19   Let $f_t(z) = \Pi_{n=1}^{\infty} (1 - tz^{2n})$, where $z \in \mathbb{C}$, $-1 \leqslant t \leqslant 1$. Study the pointwise and uniform convergence of this infinite product.

7.20   Show that if a, b, c, $d \in \mathbb{C} \backslash (\mathbb{Z} \backslash \mathbb{N})$, $a + b = c + d$, then the infinite product

$$\prod_{n=1}^{\infty} \frac{(a + n)(b + n)}{(c + n)(d + n)}$$

converges absolutely. Express its value using the gamma function.

7.21   Show that if $z = re^{it}$, $r \geqslant 0$, $t \in \mathbb{R}$, then $|\sin z| \leqslant r^{r|\sin t|}$. Moreover if $|z - k\pi| > \varepsilon$, $k \in \mathbb{Z}$, then $|\sin z| \geqslant A(\varepsilon)e^{r|\sin t|}$ for some $A(\varepsilon) > 0$.

7.22   Prove that if $f_n(z) = 1 + z/4 + \cdots + z^n/4^n$, then $|f_n(z_1) - f_n(z_2)| > |z_1 - z_2|/18$ for $|z_j| < 1$, $z_1 \neq z_2$.

7.23   Show that

$$\exp(a(z - 1/z)/2) = J_0(a) + \sum_{n=1}^{\infty} (z^n + (-z)^{-n})J_n(a)$$

where J_n is the Bessel function

$$J_n(a) = \sum_{k=0}^{\infty} \frac{(-1)^k(a/2)^{n+2k}}{k!(n+k)!}$$

7.24   Prove the Weierstrass double series theorem: Suppose that the series $\Sigma_{k=0}^{\infty} a_k^{(n)} z^k$ converge for $|z| < r$ to $g_n(z)$. If the series $\Sigma_{n=0}^{\infty} g_n(z)$ converges uniformly for $|z| < r$ to a function $G(z)$, then $G(z) = \Sigma_{k=0}^{\infty} A_k z^k$ for $|z| < r$, where $A_k = \Sigma_{n=0}^{\infty} a_k^{(n)}$, $k = 0, 1, 2, \ldots$.

7.25   An analytic version of de l'Hôpital rule: Let f and g be analytic functions in a neighborhood of z_0, not identically zero. Show that if $\lim_{z \to z_0} f(z) = \lim_{z \to z_0} g(z) = 0$, then $f(z)/g(z)$ has a limit (possibly infinite) when z approaches z_0 and $\lim_{z \to z_0} f(z)/g(z) = \lim_{z \to z_0} f'(z)/g'(z)$.

7.26   Prove the H. Cartan theorem: For arbitrary complex numbers $a_1, \ldots, a_n$ and positive H there exist disks with the sum of radii equal to $2H$ such that for any z outside the union of these disks $|(z - a_1) \cdots (z - a_n)| > (H/e)^n$.

7.27 Prove the fundamental theorem of algebra according to the scheme of d'Alembert's proof:
  (i) If a polynomial $P(z)$ would not have any root, then $|P(z)|$ attains its minimum at some z_0, $|P(z_0)| > 0$.
  (ii) If $Q(z) = P(z + z_0)/P(z_0)$, then $\min |Q(z)| = 1$.
  (iii) If $Q(z) = 1 + az^k +$ higher order terms, $a \neq 0$, then letting $z = t(-1/a)^{1/k}$, $t > 0$, observe that $|Q(z)| < 1$ for small t.

7.28 Prove the fundamental theorem of algebra using Ankeny's method:
  (i) Observe that if the polynomial $P(z) = z^{n+1} + \Sigma_{k=0}^n a_k z^k$ would not have any root, then $Q(z) = z^{n+1} + \Sigma_{k=0}^n \bar{a}_k z^k$ also vanishes nowhere.
  (ii) Apply the Cauchy integral formula to $1/PQ$ and the semicircle of radius R.
  (iii) Passing with R to infinity show that

$$\int_{-\infty}^{\infty} |P(x)|^{-2} \, dx = 0$$

7.29 Prove the Fejér-Riesz theorem: If the function f is analytic in the unit disk and continuous in its closure, then

$$2 \int_{-1}^{1} |f(x)|^2 \, dx \leqslant \int_0^{2\pi} |f(e^{it})|^2 \, dt$$

7.30 Find nontrivial solutions of the functional equation $f(2z) = 2f(z)f'(z)$.
  Show that if f is an analytic solution in a neighborhood of the origin, then f has an analytic continuation to the entire function.

7.31 Show that if λ is a complex number such that $2\lambda \notin \mathbb{Z}$, then the differential equation $zw''(z) + (2\lambda + 1)w'(z) + zw(z) = 0$ (closely related to the Bessel equation of order λ) has nontrivial entire solutions. All these solutions are proportional. Prove that they are of exponential type 1.
  Determine for arbitrary $\lambda \in \mathbb{C}$ the solution analytic in $\mathbb{C} \backslash \{0\}$ and observe that for $2\lambda \notin \mathbb{Z}$ they have removable singularities at the origin and they are mutually proportional. What happens when $2\lambda \in \mathbb{Z}$?

7.32 Prove the Looman-Menšov theorem: If f is a continuous function in a domain D with finite derivatives $\partial f/\partial x$, $\partial f/\partial y$ (except for a countable set of points) satisfying the Cauchy-Riemann equation a.e., then f is analytic in D.

7.33 The functions u and v are continuously differentiable in a domain $D \subseteq \mathbb{R}^2$. Prove that the analyticity of one of the functions $u + iv$, $u - iv$ is equivalent to the following condition: the areas of the surfaces $z = u(x, y)$, $z = v(x, y)$, $z = (u^2 + v^2)^{1/2}$, $(x, y) \in U$, are equal for every subdomain $U \subseteq D$.

7.34 Prove that if u and v are continuously differentiable functions on a domain $G \subseteq \mathbb{C}$ then either $u + iv$ or $u - iv$ is an analytic function if and only if $DuDv = 0$ and $|Du| = |Dv|$ on G.

7.35 Compute the integral $\int_{|z|=1} \exp(\sin 1/z)\, dz$.

7.36 Determine the limit

$$\lim_{n \to \infty} n^b \iint_{|z| \leq 1} \log(1 + |a_n z^n|)\, dxdy,$$

where $0 < b < 1$, $|a_n| \leq cn^p$ for some $c, p > 0$.

7.37 Find all the Laurent series of the function $\log((z - a)/(z - b))$ in the neighborhood of $z = 0$. Here log denotes the analytic branch of logarithm of $(z - a)/(z - b)$ in $\mathbb{C}\backslash[a,b]$, $a \neq b$, $ab \neq 0$.

How does one find the Laurent series of $1/\sin z$ in the annulus $\pi < |z| < 2\pi$?

7.38 Let f and g be continuous mappings from a connected subset of the complex plane into $\mathbb{C}\backslash\{0\}$. Show that the identity $f^n = g^n$ implies the relation $f = e^{2\pi i k/n}g$ for some integer k. This implication is not true when $\mathbb{C}\backslash\{0\}$ is replaced by $\mathbb{C}$.

7.39 Let f be a nonvanishing analytic function in an open subset $D \subseteq \mathbb{C}$. Prove that there exists the analytic logarithm of f on D if and only if f has the analytic roots of arbitrary order $k \geq 2$. Show that a continuous logarithm of f is automatically analytic.

7.40 Give the examples of analytic functions g on a connected open set such that:

(i) g has the analytic square root but it does not have the analytic logarithm.

(ii) g satisfies the identity $\exp(g(z)) = z$, but $g(z)$ is not of the form $\log |z| + i \arg z$ with $t < \arg z < t + 2\pi$.

7.41 Let $C_1 = \{x + iy : x \leq 1,\ y = (1 - x)/2\}$, $C_2 = \{x + i0 : x \geq 1\}$. Verify that there exist the analytic functions f and g such that $f(0) = 1$, $f^2(z) = 1 - z$ on $\mathbb{C}\backslash C_1$, $g(0) = 1$, $g^2(z) = 1 - z$ on $\mathbb{C}\backslash C_2$. Show that $f = g$ below $C = C_1 \cup C_2$ and $f = -g$ above C. What are the relations between f, g and $h(z) = \sum_{n=0}^{\infty} \binom{1/2}{n}(-z)^n$, $|z| < 1$?

7.42 Does there exist an analytic mapping f from the unit disk into itself such that $f(1/2) = 3/4$ and $f'(1/2) = 2/3$?

7.43 Show that if f is a nonvanishing analytic mapping of the unit disk, then $\sup_{|z| \leq 1/5} |f(z)|^2 \leq \inf_{|z| \leq 1/7} |f(z)|$.

7.44 Show that the analytic function f defined on the unit disk with values in the half-plane $\{z : \operatorname{Re} z > 0\}$ satisfies the estimates

$$\frac{1 - |z|}{1 + |z|} |f(0)| \leqslant |f(z)| \leqslant \frac{1 + |z|}{1 - |z|} |f(0)|$$

$$|f'(0)| \leqslant 2 |\operatorname{Re} f(0)| \leqslant 2 |f(0)|$$

7.45 Prove the Borel-Carathéodory inequality: If f is an analytic function defined on $\{z : |z| \leqslant R\}$, $M(r) = \max\{|f(z)| : |z| = r\}$, $A(r) = \max\{\operatorname{Re} f(z) : |z| = r\}$, then for all $r < R$

$$M(r) \leqslant \frac{2r}{R - r} A(R) + \frac{R + r}{R - r} |f(0)|$$

7.46 Let the analytic function f defined on the unit disk satisfy $f(0) = 1$, $|f(z)| \leqslant M$. Show that for $|z| < 1/M |f(z) - 1| \leqslant M |z|$.

Apply this to prove that if h is analytic on the unit disk, $h(0) = 0$, $h'(0) = 1$, $|h'(z)| \leqslant M$, then the image of the disk $\{z : |z| < 1/M\}$ contains the disk $\{z : |z| < 1/2M\}$.

7.47 Prove the open mapping theorem for analytic functions applying the maximum principle according to the scheme below:

Let $f(z) \neq f(a)$ for $0 < |z - a| < r$. Consider $P_b(z) = |f(z) - b|$, $|z - a| \leqslant r$, $|b - f(a)| < m/2$, where $m = \min\{|f(a + re^{it}) - f(a)| : t \in \mathbb{R}\}$. Show that $P_b(a) < \min\{|f(a + re^{it}) - b| : t \in \mathbb{R}\}$ and study z minimizing P_b.

7.48 Let f be analytic on the open set $U \setminus \{z_0\}$, z_0 being the pole of f. Show that if f is not constant on the components of U then $f(U)$ is open in $\mathbb{C} \cup \{\infty\}$.

7.49 Prove that the derivative of the injective analytic function does not vanish.

7.50 Show that if f is an analytic function and $z = a$ is a root of multiplicity k of the equation $f(z) = c$, then there exist ε, $\delta > 0$ such that for $0 < |\zeta - c| < \delta$ the equation $f(z) = \zeta$ has exactly k different roots in $\{z : 0 < |z - a| < \varepsilon\}$.

7.51 Let f be an analytic function in the unit disk and $C_r = f(\{z : |z| = r\})$ for $r \leqslant 1$. Show that if $\operatorname{diam}(C_1) \leqslant 1$, then $\operatorname{diam}(C_r) \leqslant r$.

7.52 For a function F analytic in the unit disk and $0 \leqslant u < 2\pi$ define

$$h(u) = \sup_{0 < t < 1} |F(te^{iu})|, \ g(u) = \limsup_{t \to 1 - 0} |F(te^{iu})|$$

Assume that for positive integers n $\int_0^{2\pi} h^n(u) \, du < \infty$. Show that

$$\sup_{0 \leqslant u < 2\pi} g(u) = \operatorname{ess\,sup}_{0 \leqslant u < 2\pi} g(u) = \sup_{|z| < 1} |F(z)|$$

7.53 The function w is analytic in the unit disk, $w(0) = 0$, $A \in \mathbb{C}$, $\operatorname{Re} A \geqslant 1$. Show that if $|w^2(z) + Aw(z) + zw'(z)| < 1$, then $|w(z)| < 1$.

7.54 G is a simply connected domain, symmetric with respect to the real axis, $G \neq \mathbb{C}$. An analytic function f maps G onto the unit disk, $f(a) = 0$, $f'(a) > 0$ for some $a \in G$. Show that $f(\{z \in G : \operatorname{Im} z > 0\})$ does not intersect the real axis.

7.55 An injective analytic function in the upper half-plane satisfies $\operatorname{Re} f(z) > 0$ and $f(a) = a$ for some real a. Show that $|f'(a)| \leqslant 1$.

7.56 Let an analytic injective function f map the domain G_1 onto G_2. G_1, G_2 are simply connected and different from $\mathbb{C}$. Fix $a \in G_1$.
  Show that for every injective analytic function $h : G_1 \to G_2$ such that $f(a) = h(a)$ the inequality $|h'(a)| \leqslant |f'(a)|$ holds.
  What happens if h is not injective?

7.57 Suppose that F is analytic and injective on the unit disk D, f is analytic in D and $f(D) \subseteq F(D)$. Prove that there exists an analytic function $w : D \to D$ such that $f(z) = F(w(z))$. Moreover, if $f(0) = F(0)$, then $w(0) = 0$, $|w(z)| \leqslant |z|$, $|f'(0)| \leqslant |F'(0)|$, $f(rD) \subseteq F(rD)$ for all $0 < r < 1$.

7.58 Characterize entire functions that are real on the real axis only.

7.59 Show that each analytic automorphism of the unit disk is of the form $f(z) = e^{ia}(z - z_0)/(1 - \overline{z_0}z)$ for some $a \in \mathbb{R}$, $|z_0| < 1$.

7.60 Show that if an entire function f satisfies the condition $|f(z)| = 1$ for $|z| = 1$, then $f(z) = kz^n$ for some constant k and positive integer n.

7.61 Show that if an entire function f has only real values on the unit circle, then f is constant.

7.62 Show that if an analytic map of the unit disk has two fixed points then this is the identity.

7.63 The function f is analytic in $\{z : 0 < |z| < R\}$. Determine all possible values of the integral

$$(2\pi i)^{-1} \int_{|z|=r} f'(z)(f(z) - a)^{-1}\, dz, \text{ where } a \in \mathbb{C}, \ 0 < r < R$$

Can one conclude something about the type of singularity at the origin?

7.64 Show that if ∞ is a pole of an analytic function on $\mathbb{C}$, then this function is a polynomial.

7.65 Show that if f is an injective meromorphic function on $\mathbb{C}$ (hence f has at most one singular point—a pole—in $\mathbb{C}$), then f is a homographic function.

7.66 Show that any function meromorphic on $\mathbb{C} \cup \{\infty\}$ is rational.

7.67 Suppose that f is an entire function such that $f(z + 2\pi i) = f(z)$, $|f(z)| \leqslant Ae^{B|z|}$ for some $A, B > 0$. Prove that $f = g \circ \exp$ for a function g of the form $g(z) = \Sigma_{k=-n}^{n} a_k z^k$.

7.68 Find an analytic function B bounded on the unit disk such that the function $z/(1 - z)^2 + B(z)$ maps the unit disk onto $\mathbb{C}$. ($K(z) = z/(1 - z)^2$ is the famous Koebe function).

7.69 Prove that two ellipses in the complex plane are conformally equivalent and their foci are mapped onto the foci if and only if their eccentricities are equal (and then the conformal mapping establishing the equivalence is linear).

7.70 F is an entire function satisfying the condition $\max\{|F(z)| : z \in [a,b]\} = \max\{|F(a)|, |F(b)|\}$ for every segment $[a,b]$ in $\mathbb{C}$. Prove that $F(z) = A(z - B)^n$ or $F(z) = A\exp(Bz)$ for some constants A, B.

7.71 Prove that the inverse function of an entire function is entire if and only if it is linear.

7.72 Show that any periodic entire function has a fixed point.

7.73 The entire functions f and g do not have fixed points. Show that $f \circ g$ has infinitely many fixed points.

7.74 Prove that if $f(z) = a_1 z + a_2 z^2 + \cdots$ is an entire function and $|a_1| > 1$, then the images of arbitrary neighborhood of the origin under successive iterates of f cover the complex plane with the possible exception of one point.

7.75 Show that if $P(z) = z^n + a_1 z^{n-1} + \cdots + a_n$ and $|P(z)| \leqslant 1$ for $|z| = 1$, then the polynomial P has a root on the circle $|z| = 1$.

7.76 Show that the roots of the polynomial $P(z) = a_0 z^n + a_1 z^{n-1} + \cdots + a_n$, where $0 < a_0 < \cdots < a_n$, lie outside the unit circle.

7.77 Determine the zeros of the entire function $f(z) = \Sigma_{n=0}^{\infty} A_n z^n$, where $A_0 = 1$, $A_{n+1} = rA_n/(r^{n+1} - 1)$, $|r| > 1$.

7.78 Show that if $c_1, \ldots, c_n$ are strictly positive, $c_1 + \cdots + c_n = 1$, $M = \max |a_k/c_k|^{1/k}$, then the roots of the polynomial $z^n + a_1 z^{n-1} + \cdots + a_n$ lie in the disk $\{|z| \leqslant M\}$.

7.79 Show that if P and Q are polynomials of degree at most $n - 1$ then $\max_{|z| \leqslant 1} |z^n - P(z)/Q(z)| \geqslant 1$.

7.80 Show that for arbitrarily small $r > 0$ and all sufficiently large n all zeros of the function $f_n(z) = 1 + 1/z + 1/2!z^2 + \cdots + 1/n!z^n$ lie in the disk $\{|z| < r\}$.

7.81 Prove that if $f(z) = 1 + a_1 z + a_2 z^2 + \cdots$ for $|z| \leqslant 1$ and $(2\pi\Omega)^{-1} \int_0^{2\pi} |f(e^{it})|^2 \, dt < (1 + |a_1|^2/4)^2$, then f has a zero in the unit disk.

7.82 Show that if the analytic function f has $n+1$ zeros in a convex domain, then $\operatorname{Re} f^{(n)}$ has at least one zero in this domain.

7.83 Consider a function f analytic in a bounded domain $G \subseteq \mathbb{C}$ containing the origin and continuous in $\bar{G}$. Prove that if $\operatorname{Re} \bar{z} f(z) > 0$ for $z \in bdG$ then f has exactly one zero in G.

Remark: this fact may be interpreted as a generalization of the Bolzano theorem for real functions.

7.84 Prove the Darboux mean value theorem for complex integrals: If $g(z)$ is continuous on the segment $[a,b] \subseteq \mathbb{C}$, $|a| < |b|$ and $f(r)$ is a real positive function defined for $|a| < r < |b|$, $r = |z|$, then

$$\int_a^b f(r)g(z) \, dz = cg(z_0) \int_a^b f(r) \, dz$$

for some $|c| \leqslant 1$ and z_0 belonging to the segment $[a,b]$.

7.85 Prove the local mean value theorem for analytic functions: If f is an analytic function in a domain containing z, then there exists a neighborhood U of z such that for every $z_1 \in U$ there is z_2 in the disk of radius $|z - z_1|/2$ centered at $(z + z_1)/2$ such that $f(z_1) - f(z) = (z - z_1)f'(z_2)$.

7.86 Prove the Riemann-Macdonald theorem: If a function f is analytic inside and on a simple closed curve C, $|f(z)| = \text{const}$, $f'(z) \neq 0$ on C and if f has k zeros inside C, then f' has $k - 1$ zeros there.

7.87 Study the roots of the equation $\tan z = az$ with the parameter $a > 0$.

7.88 How many complex roots has the equation $\sin z = z$?

7.89 For an entire function f and $r > 0$ define $M(r) = \sup_{|z|=r} |f(z)|$, $n(r) =$ number of zeros of f in the disk $\{z : |z| < r\}$ (counting multiplicities). Show that if $f(0) = 1$, then $n(r) \log 2 \leqslant \log M(2r)$.

7.90 Prove that, in the notation of preceding problem, $\int_0^r n(t)t^{-1} \, dt = \sum_{k=1}^m \log(r/|z_k|)$, where $z_1, \ldots, z_m$ denote the zeros of f in the disk $\{z : |z| < r\}$.

7.91 Verify that the function

$$f(z) = \sum_{n=1}^{\infty} \exp(-n^2) \sum_{m=1}^{n-1}{}' \frac{1}{z - m/n}$$

where $'$ denotes the summation over m, n relatively prime, is analytic in $\mathbb{C}\setminus[0,1]$, and $\lim_{y \to 0} |f(r + iy)| = \infty$ for all rational $r \in (0,1)$.

7.92 Does there exist an entire transcendental function which tends to infinity along all half-lines starting from the origin?

7.93 Show that the entire function defined by the series $\sum_{n=0}^{\infty} z^n/\Gamma(n/2 + 1)\Gamma(n + 1)$ tends to zero faster than any power of $1/|z|$ if z tends to infinity in any sector $3\pi/4 + \varepsilon < \arg z < 5\pi/4 - \varepsilon$, $\varepsilon > 0$.

7.94 Does there exist a nonzero entire function which tends to zero along all half-lines starting from the origin?

7.95 Does there exist a nonzero entire function such that all its integrals along the lines passing through the origin vanish?

7.96 For an entire function $f(z) = \sum_{n=0}^{\infty} a_n z^n$ and $r > 0$ define $m(r) = \sup\{|a_n|r^n : n \in \mathbb{N}\}$, $M(r) = \sup\{|f(z)| : |z| = r\}$. Suppose that for some $a, b > 0$ $m(r) \leqslant a \exp(r^b)$. Prove that $M(r) \leqslant Ar^b m(r) + B$ for some A, B.

7.97 Show that if g is analytic in $\{z : |z| > r\}$ and C is a rectifiable closed curve in this set, then $f(z) = \oint_C e^{zw} g(w)\, dw$ is of exponential type $\leqslant r$, that is, $|f(z)| \leqslant e^{(r+\varepsilon)|z|}$ for every $\varepsilon > 0$ and sufficiently large $|z|$.

Prove that if an entire function f is of exponential type $a < b < \infty$, then $a_k = f^{(k)}(0)/k!$ satisfy the condition $|a_k| \leqslant (be/k)^k$ for large k. Further prove that the radius of convergence R of $F(z) = \sum_{k=0}^{\infty} k! a_k z^k$ is at least $1/(ae)$ and $f(z) = (2\pi i)^{-1} \oint_{|w| = r} e^{zw} w^{-1} F(w^{-1})\, dw$, where $1/R < r < \infty$. What is F if f is $\exp z$, $\cos z$?

7.98 Construct an entire function f such that $f(x) \geqslant F(x)$ for given increasing function $F : \mathbb{R}^+ \to \mathbb{R}^+$ and all $x \geqslant 2$.

7.99 Show that if ∞ is an isolated singularity of the function f, then the limit $\lim_{r \to \infty} (\log(M(r))^+/\log r$ exists, $M(r) = \sup\{|f(z)| : |z| = r\}$.

Prove that the limit above is 0 if and only if ∞ is a removable singularity, it is equal to n if f has a pole of order n at ∞, and it is infinite if ∞ is an essential singularity of f.

7.100 Prove the "intermediate" Picard theorem for entire functions of moderate growth: Let $\log_1 t = \log t$, $\log_{n+1} t = \log(\log_n t)$. If f is a nonconstant entire function such that

$$\limsup_{r \to \infty} \log_n \left(\max_{|z| = r} |f(z)| \right) \Big/ \log r < \infty$$

and $f(z) \neq a$ for some $a \in \mathbb{C}$, then f takes on every other complex value $b \neq a$ infinitely often.

7.101 Does there exist a real analytic function f such that the Taylor series of f at $x = N$ converges in the disk of radius exactly equal to $1/N$?

7.102 For a function $f : [-1, 1] \to \mathbb{R}$ and $n \in \mathbb{N}$ define $m_n = \inf\{\|f - P\|_\infty :$ P is a polynomial of degree $\leqslant n\}$. Show that f is the restriction of a function analytic in some neighborhood of $[-1, 1]$ if and only if $\limsup_{n \to \infty} n^{-1} \log m_n < 0$.

7.103 Does there exist an analytic function in the unit disk such that $|f(1/n)| \leqslant e^{-n}$?

7.104 Prove that each bounded analytic function in the half-plane $\{z : \operatorname{Re} z > 0\}$ which vanishes for $z = c_n > 0$ with $\Sigma_{n=0}^\infty 1/c_n = \infty$ is identically zero.

7.105 Prove the Carleman theorem: The function g is analytic in the sector $S = \{z : 0 < |z| < B, \; |\arg z| < \pi/2\}$ and continuous in its closure. If $|g(z)| \leqslant b_n |z|^n$ in S for all n and $\Sigma_{n=0}^\infty b_n^{-1/n} = \infty$, then g is identically zero.

7.106 Show that given a complex sequence (a_n) there exists at most one function f analytic in the half-plane $\{z : \operatorname{Re} z > 0\}$ such that $f(n) = a_n$, $n = 1, 2, \ldots, |f(z)| \leqslant C_1 \exp(C_2 |z|)$ for some constants $C_1 > 0, 0 < C_2 < \pi$.

7.107 Prove that the series $\Sigma_{n=1}^\infty (-1)^{[\sqrt{n}]} z^n / n$ is conditionally convergent on the unit circle.

7.108 Prove that if $\lim_{n \to \infty} a_n = 0$, $\Sigma_{n=1}^\infty |a_n - a_{n-1}| < \infty$, then the power series $\Sigma_{n=0}^\infty a_n z^n$ converges on the unit disk. Show that if its radius of convergence is one, then this series converges on the circle $|z| = 1$, possibly except for $z = 1$.

7.109 Prove the following extension of the Pringsheim theorem: If $|\arg a_n| \leqslant c < \pi/2$ and the radius of convergence of the series $f(z) = \Sigma_{n=0}^\infty a_n z^n$ is $R, 0 < R < \infty$, then $z = R$ is a singularity of f.

7.110 Show that $f(z) = \Sigma_{n=0}^\infty z^{2^n}$ and, more generally, the Weierstrass function $\Sigma_{n=0}^\infty a^n z^{b^n}$, where $b > 1$ is an integer, cannot be continued beyond the unit disk (which is the original disk of convergence).

7.111 Show that $z = 1$ is a singular point of the analytic function $f(z) = \Sigma_{n=0}^\infty z^{2^n} / n!$. However the limit of the derivative $f^{(k)}(z)$ exists if z tends to 1 for all k. Prove that f has no analytic continuation beyond the unit disk.

7.112 Prove the Kronecker theorem: The power series $\Sigma_{n=0}^\infty a_n z^n$ represents a quotient of polynomials if and only if the determinant $\det(a_{i+j})$, $i, j = 0, \ldots, n$, vanishes for all sufficiently large n.

7.113 Prove the Szegö theorem: If the sequence (a_n) of the coefficients of the power series $f(z) = \Sigma_{n=0}^\infty a_n z^n$ contains finitely many different numbers, then either $f(z) = P(z)/(1 - z^m)$ for some polynomial P, or f has no analytic continuation beyond its disk of convergence.

7.114 Prove that the Fredholm series $\sum_{n=0}^{\infty} a_n z^{n^2}$, $0 < a < 1$, represents the analytic function with no analytic continuation beyond the unit disk. However $f(e^{it})$ is a smooth function of t.

7.115 Consider a construction of a Cantor set in the square: Divide the square $[0, 1] \times [0, 1]$ into nine squares and omit five squares forming a cross in which either x or y or both lie between $1/3$ and $2/3$, limits excluded. The remaining four squares are treated in the same manner and so on. After n subdivisions and omissions there are left 4^n squares of total area $(4/9)^n$. The perfect set C is obtained as the intersection of the remaining closed squares. Denote the centers of the squares by $z_{n,k}$, $k = 1, \ldots, 4^n$, and define $f_n(z) = 4^{-n} \sum_{k=1}^{4^n} (z - z_{n,k})^{-1}$.

Show that $\lim_{n \to \infty} f_n(z) = f(z)$ is continuous in $\mathbb{C} \cup \{\infty\}$, and analytic outside C, $f(\infty) = 0$ and $\oint_{|z|=2} f(z)\,dz = 2\pi i$.

7.116 The polynomial series $\sum_{n=0}^{\infty} (z(1 - z))^{3^n}$ converges in the domain bounded by the lemniscate $|z(1 - z)| = 1$ and it has no analytic continuation outside this domain. If each term is expanded in powers of z and the resulting expressions are written consecutively, one obtains the power series of $f(z)$ converging in the unit disk only.

Show that there exists a subsequence of the partial sums of this power series converging beyond its disk of convergence. This is a phenomenon of overconvergence.

7.117 Prove the Fatou-Pólya theorem: Given a power series $\sum_{n=0}^{\infty} a_n z^n$ of the convergence radius equal to 1 there exists a choice of signs $\varepsilon_n \in \{-1, 1\}$ such that the function $\sum_{n=0}^{\infty} \varepsilon_n a_n z^n$ is not continuable beyond the unit disk.

7.118 Consider the series

$$\sum_{n=1}^{\infty} \frac{1}{2^{n+1}} \left(1 - \sqrt{\left(1 - \frac{z}{2n}\right)\left(1 - \frac{z}{2n + 1}\right)}\right)$$

where the branches of the square root are chosen independently in each term. So for $z = 0$ one obtains the series $\sum_{n=0}^{\infty} \varepsilon_n 2^{-n}$, where $\varepsilon_n = 0$ or 1 according to the choice of the branch. In other words these series correspond to the reals in $(0, 1)$. Describe the equivalence classes of this series with respect to the relation "have a common analytic continuation."

7.119 Consider the polynomials $g_m(z) = (1 - z^m)/(1 - z)$ and $h_m(z) = \sum_{k=0}^{m-1} z^{km} g_m(e^{-2\pi k i'm} z)$. Prove that the Luzin power series obtained from the polynomial series $\sum_{m=1}^{\infty} z^{1^2 + \cdots + (m-1)^2} h_m(z) m^{-1/2}$ has the coefficients converging to zero and it diverges for $|z| = 1$.

Show that if a_n denote the absolute values of the coefficients of the power series above, then the Sierpiński series $a_0 - a_0 z + a_1 z^2 - a_1 z^3 + a_2 z^4 - a_2 z^5 + \cdots$ converges on the unit circle only for $z = 1$.

7.120 Prove that if $a_1 > 0$, $a_{n+1} > a_n + \delta$ for some $\delta > 0$, $\lim_{n \to \infty} n/a_n = s$, then the Dirichlet series $\Sigma_{n=1}^\infty c_n \exp(a_n z)$ with the coefficients satisfying the relation $\limsup_{n \to \infty} (\log|c_n|)/a_n = c$ has a singularity on each segment of the form $\operatorname{Re} z = c$, $b < \operatorname{Im} z < b + a$, $a > 2\pi s$.

7.121 Prove the Mittag-Leffler theorem: Given any domain D in the complex plane there exists an analytic function on D with no analytic continuation in any neighborhood of any point of $bd|D$ (that is, a function whose holomorphy domain is D).

7.122 For a function f analytic in a neighborhood of the origin define $A(f) = \{a \in \mathbb{C} : f \text{ has an analytic continuation along the segment } [0, a]\}$.

The Hadamard theorem ([Hi1], Th. 11.6.1) states that if $f(z) = \Sigma_{n=0}^\infty a_n z^n$, $g(z) = \Sigma_{n=0}^\infty b_n z^n$, $h(z) = \Sigma_{n=0}^\infty a_n b_n z^n$ in a neighborhood of the origin then $A(h) \supseteq (A(f)^c \cdot A(g)^c)^c$, ($E^c$ denotes the complement of E, $\cdot$ denotes multiplication).

Give the examples where this inclusion is strict.

The multiplicative Hadamard theorem (in the simplest version) says that the singularities of the power series $\Sigma_{n=0}^\infty a_n b_n z^n$ lying in the disk of radius $R_a R_b$ are of the form ab, where a and b are the singular points of the series $\Sigma_{n=0}^\infty a_n z^n$, $\Sigma_{n=0}^\infty b_n z^n$ respectively and R_a, R_b are their radii of convergence.

Show that the series $\Sigma_{n=0}^\infty a_n b_n z^n$ may converge in a disk of radius strictly greater than $R_a R_b$.

7.123 Prove the Borel theorem: If, in the notation of the preceding problem, a and b are the poles of f and g respectively and $z = ab$ is the unique representation of z as the product of numbers from $A(f)^c$, $A(g)^c$, then z is a pole of h.

7.124 Prove the Faber theorem: If, in the notation of the problem 7.122, a and b are the isolated singular points of f and g respectively, one of them being an essential singularity and $z = ab$ is the unique representation of z as the product of numbers from $A(f)^c$, $A(g)^c$, then z is an essential singularity of h.

7.125 Let $f(z) = \Sigma_{n=1}^\infty a_n z^n$, where a_n lie inside a Jordan curve, the origin being outside. Prove that there exist singularities $z_1, \ldots, z_k$ of f on the unit circle such that $z_1^{n_1} \cdots z_k^{n_k} = 1$ for some positive integers $n_1, \ldots, n_k$.

7.126 Prove the Dieudonné theorem: If the function $f(z) = z + \Sigma_{n=2}^\infty c_n z^n$ with real c_n is analytic and injective in the unit disk, then $|c_n| \leqslant n$.

7.127 Prove the Lindelöf theorem: Suppose that f is analytic and bounded in the unit disk D and the continuous curve $C : [0, 1) \to D$ tends to 1 when t tends to 1. Show that if $\lim_{t \to 1} f(C(t)) = L$, then the limit of f at $z = 1$ along each curve nontangential to the circle is equal to L.

7.128 Prove the generalized Helmer theorem: If the entire functions f and g have no common zeros, then there exist entire functions a and b, $a(z) \neq 0$ for all z, such that $af + bg \equiv 1$.

If f and g are nonconstant polynomials and a and b vanish nowhere, then a and b are constant.

7.129 Prove the Borel lemma: The equation $\exp(f(z)) + \exp(g(z)) \equiv 1$ has no nonconstant entire solutions f and g.

More generally, show that if $\exp(f_1(z)) + \cdots + \exp(f_n(z)) \equiv 1$ for entire functions $f_1, \ldots, f_n$, then some f_j must be constant.

7.130 Prove the Vitali-Porter-Lindelöf theorem: Suppose that analytic functions f_n, $n = 1, 2, \ldots$, are uniformly bounded in a domain $D \subseteq \mathbb{C}$ and for some sequence $(z_n) \subseteq \mathbb{C}$ having a cluster point in D the limits $\lim_{n \to \infty} f_n(z_k)$ exist for all k. Prove that $f(z) = \lim_{n \to \infty} f_n(z)$ exists for every $z \in D$, f is the analytic function and f_n converge to f uniformly on every compact in D.

7.131 Show that assuming in the preceding problem the pointwise convergence of f_n instead of the uniform boundedness of (f_n), the limit function f is analytic on an open dense subset of D. Give an example where this set is not equal to D.

7.132 Given meromorphic functions $f_1, \ldots, f_n$ in a domain $D \subseteq \mathbb{C}$ does there exist an analytic function g in D such that $f_j(z) \neq g(z)$ for $j = 1, 2, \ldots, n$ and all $z \in D$?

7.133 The functions f_n are analytic in a simply connected domain D, M is the closure of the linear subspace spanned by (f_n) in the topology of the uniform convergence on compacts. Show that if $g \notin M$, then there exists a functional $\ell(f) = \oint_C f(t)w(t)\,dt$, where C is a closed curve in D, the function w is analytic on and outside C, $w(\infty) = 0$, such that $\ell(f_n) = 0$ and $\ell(g) \neq 0$.

7.134 Prove the Markuševič theorem: The system (f_n) is complete in the space of analytic functions on a simply connected domain D with the topology of uniform convergence on compacts if and only if for every function w analytic on and outside the closed curve $C \subseteq D$ such that $w(\infty) = 0$, the equality $\oint_C f_n(t)w(t)\,dt = 0$ for all n implies $w \equiv 0$.

7.135 Show that if $f(z) = \sum_{n=0}^{\infty} a_n z^n$, $a_n \neq 0$, is an entire function and b_n are different complex numbers, the sequence (b_n) being bounded, then the system $(f(b_n z))$ is complete in the space of entire functions with the topology of uniform convergence on compacts.

7.136 Prove the Gelfond theorem: If $f(z) = \sum_{n=0}^{\infty} a_n z^n$, $a_n \neq 0$, is an entire function of order ρ, $0 < \rho < \infty$ (that is, ρ is the minimal number such that $|f(z)| < \exp(|z|^{\rho + \varepsilon})$ for every $\varepsilon > 0$ and all sufficiently large z) and $\tau > \rho$ is the exponent of convergence of the complex sequence (c_n) (that is, for every $\varepsilon > 0$

$\Sigma_{n=0}^{\infty} |c_n|^{-(\tau+\varepsilon)} < \infty$ but $\Sigma_{n=0}^{\infty} |c_n|^{-\tau+\varepsilon} = \infty$), then the system of functions $(f(c_n z))$ is complete in the space of entire functions with the topology of uniform convergence on compacts.

7.137 Prove the Müntz-Szász theorem: If positive reals c_n strictly increase to infinity, $\Sigma_{n=1}^{\infty} 1/c_n = \infty$, then the span of the functions 1, (x^{c_n}) is dense in $C[0, 1]$.

7.138 Construct an entire function f such that the set of derivatives $\{f^{(n)} : n = 1, 2, \ldots\}$ is dense in the space of entire functions with the topology of uniform convergence on compacts.

7.139 The set of all analytic functions on the unit disk with the topology of uniform convergence on compacts is a locally convex topological vector space. Let K denote the set of the functions $f(z) = z + \Sigma_{n=2}^{\infty} a_n z^n$ which map the unit disk onto a convex domain. Show that for $k \geqslant 2 |z + a_k z^k \in \overline{\text{conv}} \, K$ if and only if $|a_k| \leqslant 1/2$.

7.140 Consider the map $F : \{z : |z| = 1\}^n \to \mathbb{C}^m$ given by the formula $F(z_1, \ldots, z_n) = (w_1, \ldots, w_m)$ where $w_k = z_1^k + \cdots + z_n^k$, $k = 1, \ldots, m$. Prove that for every m, $R > 0$ and sufficiently large n the range of F contains the polydisk $\{(w_1, \ldots, w_m) : |w_j| \leqslant R\}$.

7.141 Prove the Forelli theorem: If a complex function f defined in a neighborhood U of the origin in $\mathbb{C}^n$ is analytic on the intersection of U with each line $\{cz : c \in \mathbb{C}\}$ and f is C^∞ in U as a real function, then f is holomorphic in U.

Give the example which shows that the condition $f \in C^\infty$ is important.

Compare this with the classical Hartogs theorem.

7.142 Prove that if $F : G \to \mathbb{C}^n$ is a holomorphic injective map defined on the open set G in $\mathbb{C}^n$, then the jacobian of F does not vanish; hence F is biholomorphic.

7.143 Prove the following version of the Paley-Wiener theorem: If C is a balanced convex set in $\mathbb{R}^n$, $C^o = \{k : (k, x) \geqslant -1$ for all $x \in C\}$ and $\rho(v) = \sup\{(v, x) : x \in C\} = \inf\{a > 0 : av \in C^o\}$, $v \in \mathbb{R}^n$, is its gauge (Minkowski functional of C^o), then the support of the function $f \in \mathscr{S}(\mathbb{R}^n)$ is in C if and only if $\hat{f}$ is the restriction to $\mathbb{R}^n$ of an entire function $\tilde{f}$ satisfying for each m and some constants D_m the estimate

$$|\tilde{f}(z)| \leqslant D_m (1 + |z|^2)^{-m} e^{-\rho(\operatorname{Im} z)}$$

7.144 Suppose that f is an entire function in $\mathbb{C}^n$ such that for every $\varepsilon > 0$ there exist constants $N = N(\varepsilon)$, $C = C(\varepsilon)$ satisfying the condition

$$|f(z)| \leqslant C(1 + |z|)^N e^{\varepsilon |\operatorname{Im} z|} \qquad z \in \mathbb{C}$$

Prove that f is a polynomial.

7.145 The entire function f in $\mathbb{C}^n$ satisfies the estimate $|f(z)| \leqslant (1 + |z|)^N e^{r|\operatorname{Im} z|}$ for some $N, r > 0$ and $|f(x)| \leqslant 1$ for all $x \in \mathbb{R}^n$. Prove that for all $z \in \mathbb{C}^n$ $|f(z)| \leqslant e^{r|\operatorname{Im} z|}$.

7.146 Construct examples of nonlinear holomorphic automorphisms of $\mathbb{C}^n$, $n > 1$.

7.147 It is well-known that each invertible analytic function in $\mathbb{C}$ is a surjection (hence linear, see Problem 7.71). This is not the case for several complex variables which follows from the analysis of the Fatou example: For $0 < a < 1$ $A(z_1, z_2) = (-az_1, az_2)$ and $T(z_1, z_2) = (z_2, a^2 z_1 + (1 - a^2) z_2^2)$ are the automorphisms of $\mathbb{C}^2$. The functional equation $T \circ F = F \circ A$ has a biholomorphic solution F such that the range $F(\mathbb{C}^2)$ is not all space $\mathbb{C}^2$.

Verify this statement.

7.148 Prove the H. Cartan uniqueness theorem: If F is the holomorphic map of a bounded domain in $\mathbb{C}^n$ into itself and there exists x_0 such that $F(x_0) = x_0$, $DF(x_0) = Id$, then F is the identity.

7.149 Consider domains G_1, G_2 in $\mathbb{C}^n$ satisfying the following conditions: $0 \in G_j$ and if $z \in G_j$ then $e^{it} z \in G_j$ for all $t \in \mathbb{R}$, $j = 1, 2$. F is the holomorphic map from the bounded domain G_1 onto G_2 such that $F(0) = 0$. Show that F is linear.

This fact is not true without the boundedness assumption on G_1; give an example.

7.150 Construct a holomorphic curve $C : \mathbb{C} \to \mathbb{C}^n$ with dense range.

7.151 F is an entire function on $\mathbb{C}$ such that on the set $A = \{z = (z_1, \ldots, z_n): F(z) = 0\}$ the estimate $|z_n| \leqslant C(1 + |(z_1, \ldots, z_{n-1})|^m)$ holds for some constants C, m.

Prove that A is the set of zeros of a polynomial.

8
Fourier Series

Topics covered include:
trigonometric polynomials, convergence of Fourier series, general orthogonal series, and Fourier transform.

8.1 Show that for integers $p < q$ and real $x \neq 0$ (mod 2π) $|\Sigma_{n=p}^{q} e^{inx}| \leq 1/|\sin x/2|$. Moreover show that if $c_p \geq c_{p+1} \geq \cdots \geq c_q \geq 0$, then $|\Sigma_{n=p}^{q} c_n e^{inx}| \leq c_p/|\sin x/2|$.

8.2 Prove that for $c_n \geq c_{n+1}$, $\lim_{n \to \infty} c_n = 0$ the series $\Sigma_{n=0}^{\infty} c_n e^{inx}$ converges for all $x \neq 0$ (mod 2π) and this convergence is uniform on compacts which do not contain any integer multiple of 2π.

8.3 Show the estimate $|\Sigma_{n=1}^{N} c_n \sin nx| \leq A(\pi + 1)$ if $c_n \geq c_{n+1} \geq 0$, $nc_n \leq A$ for some $A > 0$.

Using the formula $(\pi - x)/2 = \Sigma_{k=1}^{\infty} k^{-1} \sin kx$, $0 < x < 2\pi$, show that for all x and n

$$\left| \frac{\sin x}{1} + \frac{\sin 2x}{2} + \cdots + \frac{\sin nx}{n} \right| \leq \frac{\pi}{2} + 1$$

8.4   Prove that if $c_n \geqslant c_{n+1}$, $\lim_{n\to\infty} c_n = 0$, then the series $\Sigma_{n=1}^{\infty} c_n \cos nx$ uniformly converges if and only if $\Sigma_{n=1}^{\infty} c_n < \infty$ and the series $\Sigma_{n=1}^{\infty} c_n \sin nx$ uniformly converges if and only if $\lim_{n\to\infty} nc_n = 0$.

8.5   Show that if $|a_1 \sin x + a_2 \sin 2x + \cdots + a_n \sin nx| \leqslant |\sin x|$, then $|a_1 + 2a_2 + \cdots + na_n| \leqslant 1$.

8.6   Show that each trigonometric polynomial of degree N has at most $2N$ roots in $[0, 2\pi)$ (counting their multiplicities).

8.7   The function f is a real trigonometric polynomial of degree at most N, $\sup |f(x)| = f(x_0)$. Prove the Stečkin lemma: $f(x_0 + y) \geqslant f(x_0) \cos Ny$ for $|y| \leqslant \pi/N$.

8.8   Prove the Bernstein inequality: $\|f'\|_\infty \leqslant N \|f\|_\infty$ valid for all trigonometric polynomials of degree less than or equal to N.

8.9   Show that any positive trigonometric polynomial f is of the form $f = |g|^2$ for some trigonometric polynomial g.

8.10   Prove that for every finite subset $F \subseteq \mathbb{Z}$ and every $\varepsilon > 0$ there exists a trigonometric polynomial f such that $0 \leqslant \hat{f}(n) \leqslant 1$ for all $n \in \mathbb{Z}$, $\hat{f}(n) = 1$ for $n \in F$ and $\|f\|_1 \leqslant 1 + \varepsilon$.
   What subsets F admit $\varepsilon = 0$?

8.11   Show that if a real trigonometric polynomial $f(x) = \Sigma_{k=1}^{n}$ $(a_k \sin kx + b_k \cos kx)$ satisfies the conditions $|f(x)| \leqslant 1$ for $0 \leqslant x \leqslant 2\pi$, $|f(x_j)| = 1$ for $0 \leqslant x_1 < x_2 < \cdots < x_{2n} < 2\pi$, then $f(x) = \cos(nx + a)$ for some a.

8.12   Let $\mathscr{E}$ denote the set of all trigonometric polynomials $f(x) = c_0 + c_1 \cos x + \cdots + c_n \cos nx$ with $c_0 \geqslant c_1 \geqslant \cdots \geqslant c_n \geqslant 0$. Show that $1 - 2/\pi \leqslant (n+1) \min_{f \in \mathscr{E}} (\max_{\pi/2 \leqslant x \leqslant \pi} |f(x)|/\max_{0 \leqslant x \leqslant \pi} |f(x)|) \leqslant 2^{-1} + 2^{-1/2}$.

8.13   Assume that the trigonometric polynomial $f(x) = \Sigma_{k=1}^{n} a_k \cos kx$ decreases on $[0, \pi]$. Show that the trigonometric polynomial $f(x) = \Sigma_{k=1}^{n} a_k \sin kx$ is nonnegative in the same interval.

8.14   Prove that

$$\sum_{n=1}^{\infty} \frac{\sin n}{n} = \sum_{n=1}^{\infty} \left(\frac{\sin n}{n}\right)^2 = \frac{\pi - 1}{2} \qquad \sum_{n=1}^{\infty} \frac{\sin^2 n}{n^4} = \frac{(\pi - 1)^2}{6}$$

8.15   Does the series $\Sigma_{n=2}^{\infty} \sin nt/(n \log n)$, $0 < t < \pi$, converge absolutely?

8.16   Is the series $\Sigma_{n=-\infty}^{\infty} (-1)^n e^{inx}((1 + |n|) \log(2 + n^2))^{-1}$ the Fourier series of a bounded function?

8.17 Prove the formula $\sin^{-2} t = \sum_{n=-\infty}^{\infty} (t - n\pi)^{-2}$ using the Parseval identity.

8.18 Using the identity

$$\frac{1}{\pi} \int_0^\pi (\cos(a+n)x + \cos(a-n)x)\, dx = (-1)^n \frac{2a}{a^2-n^2} \frac{\sin a\pi}{\pi} \qquad a \in \mathbb{R}\backslash\mathbb{Z}$$

prove that

$$\frac{\pi}{2} \frac{\cos ax}{\sin a\pi} = \frac{1}{a\pi} + \sum_{n=1}^{\infty} \frac{a \cos nx}{a^2 - n^2}$$

Put in this identity $x = 0$, $t = a\pi$ and compute the integral $\int_0^\infty \frac{\sin x}{x}\, dx$ decomposing it into a series of integrals.

8.19 Prove the Fejér lemma: For all $f \in L^p(0, 2\pi)$, $g \in L^q(0, 2\pi)$, $1/p + 1/q = 1$, $1 \le p \le \infty$

$$\lim_{|n| \to \infty} \int_0^{2\pi} f(t)g(nt)\, dt = 2\pi \hat{f}(0)\hat{g}(0)$$

8.20 Prove that for $f \in L^1(0, 2\pi)$, $g \in L^\infty(0, 2\pi)$

$$\lim_{r \to 1-0} \sum_{n=-\infty}^{\infty} r^{|n|} \hat{f}(n)\hat{g}(-n) = \frac{1}{2\pi} \int_0^{2\pi} f(t)g(t)\, dt$$

$$= \lim_{N \to \infty} \sum_{n=-N}^{N} \left(1 - \frac{|n|}{(N+1)}\right) \hat{f}(n)\hat{g}(-n)$$

If $f \in L \log^+ L$, $g \in L^\infty$, then $\lim_{N \to \infty} \sum_{n=-N}^{N} \hat{f}(n)\hat{g}(-n) = \frac{1}{2\pi} \int_0^{2\pi} f(t)g(t)\, dt$.

8.21 Prove the Mercer theorem generalizing the classical Riemann-Lebesgue lemma: If (f_n) is an orthonormal system in L^2 on $(0, 1)$ uniformly bounded in L^∞, then the Fourier coefficients of any L^1 function with respect to the system (f_n) tend to zero.

8.22 Prove the Cantor-Lebesgue lemma: If $(a_n \cos nx + b_n \sin nx)$ converges to zero in a set of positive measure, then $\lim_{n \to \infty} a_n = \lim_{n \to \infty} b_n = 0$.

8.23 Prove that the system of functions $\{\sin(2^n \pi x), n \ge 0\}$, is not linearly dense in $L^2(A, m)$ for any subset $A \subseteq [0, 1]$ of strictly positive measure.

8.24 Is it true that for each measurable set $E \subseteq (0, 1)$ the numbers $C_n = \int_E \exp(2\pi i x)\, dx$ satisfy the inequality $|nC_n| \le M$ for some constant $M = M(E)$?

8.25 Prove the Denjoy-Luzin theorem: If a trigonometric series $\sum_{n=1}^{\infty} (a_n \cos nx + b_n \sin nx)$ converges absolutely on a set of positive measure, then $\sum_{n=1}^{\infty} (|a_n| + |b_n|) < \infty$.

8.26 The sequence $(\cos nx)$, $n \geqslant 0$, has the following property: any of its terms cannot be approximated uniformly on $[0, \pi]$ by linear combinations of the other terms. Is it true for an interval $[0, a]$, $a < \pi$?

8.27 Show that if f is integrable on $(0, 1)$ and $\int_0^1 f(t)e^{i\pi t}\, dt = 0$ for all $n \geqslant N$, then $f = 0$ a.e. Can one replace the interval $(0, 1)$ by a greater one?

8.28 Show that if f is integrable on $(0, 2\pi)$ and $\int_0^{2\pi} f(t) \exp(ia_n t)\, dt = 0$, $n = 1, 2, \ldots$, for a complex sequence (a_n) having at least one cluster point, then $f = 0$ a.e.

8.29 Prove that if the function f is of bounded variation on $[0, 2\pi)$, then $|n\hat{f}(n)| \leqslant V(f, [0, 2\pi))/2\pi$ for all integers n.

8.30 Assume that $m \leqslant f(x) \leqslant M$ on $[0, 2\pi)$ and that the Fourier coefficients of the function f satisfy the estimates $|a_n| \leqslant A/n$, $|b_n| \leqslant B/n$, $n > 0$. Show that the partial sums $s_n f$ of the Fourier series of f satisfy the inequalities $m - A - B \leqslant s_n f \leqslant M + A + B$.

8.31 Denote for a smooth 2π-periodic function f $M_k = \|D^k f\|_1$, $k = 1, 2, \ldots$. Show that for nonzero integers n $|\hat{f}(n)| \leqslant M_k |n|^{-k}$. Show that if f is in a Gevrey class, that is, $\|D^k f\|_\infty \leqslant Cb^k(k!)^a$ for some $C, b > 0$, $a > 1$ and all k, then $|\hat{f}(n)| \leqslant C' \exp(-b'|n|^{1/a})$ for some constants $b', C' > 0$.

8.32 Show that if the Fourier coefficients of a function f satisfy the condition $\sum_{n=-\infty}^{\infty} |n|^c |\hat{f}(n)| < \infty$ for some $c > 0$, then f is differentiable $[c]$ times and its derivative of order $[c]$ satisfies the (uniform) Hölder condition with exponent $c - [c]$.

8.33 Prove that for a 2π-periodic function f of class C^k $\|s_n f - f\|_\infty \leqslant Cn^{-k+1/2}$ with some $C = C(f)$. Show that this inequality holds for $k = 1$ even if f is continuous and only piecewise C^1.

8.34 Define for a 2π-periodic integrable function f the integral modulus of continuity $\omega_1 f(h) = \|T_h f - f\|_1 = \omega_1 f(-h)$, where T_h is the translation by h. Show that $|\hat{f}(n)| \leqslant \omega_1 f(\pi/n)/2$ for nonzero integers n.
  Show that if $(\int_0^{2\pi} |f(t + h) - f(t)|^p\, dt)^{1/p} = O(|h|^a)$, $0 < a < 1$, then $\hat{f}(n) = O(|n|^{-a})$.

8.35 Let f be a 2π-periodic integrable function. Prove that the following conditions are equivalent:
  (i) $\hat{f}(n) = O(e^{-\varepsilon|n|})$ for some $\varepsilon > 0$.

(ii) f is a.e. the restriction to $\mathbb{R}$ of an analytic function in the strip $|\text{Im } z| < \delta$ for some $\delta > 0$.

(iii) The Fourier series $\Sigma_{n=-\infty}^{\infty} \hat{f}(n)e^{inz}$ converges uniformly in the strip $|\text{Im } z| < \delta$ for some $\delta > 0$.

(iv) The series above converges for some $z = z_1$, $z = z_2$ with $\text{Im } z_1 > 0$, $\text{Im } z_2 < 0$.

8.36 Show that for the bounded function f $\|s_n f\|_\infty = O(\log n)$.

8.37 Prove that for the continuous function $f \in C(\mathbb{T})$ $\|s_n f\|_\infty = o(\log n)$.

8.38 Prove that for a 2π-periodic integrable function f which is differentiable at $x_0 \in (-\pi, \pi)$ the (nonsymmetric) partial Fourier sums $s_{m,n}(x_0) = \Sigma_{k=-m}^{n} \hat{f}(k) \exp(ikx_0)$ converge to $f(x_0)$ when m, n tend to infinity.
This result can be generalized easily to the case of a function satisfying the one-sided Hölder condition at the point of discontinuity.

8.39 Show that the linear operator defined for periodic continuous functions f on $[0, 2\pi]$ by $Tf(x) = \Sigma_{n=-\infty}^{\infty} \hat{f}(n)e^{inx}$ is not bounded in $C(\mathbb{T})$.

8.40 Prove the Fatou theorem: If f is integrable and if $\lim_{N\to\infty} N^{-1} \Sigma_{n=-N}^{N} |n\hat{f}(n)| = 0$, then $\lim_{N\to\infty} s_N f(x) = f(x)$ a.e. Moreover if f is continuous, then $s_N f$ converges uniformly.

8.41 Show that if a sequence (c_n), $n \in \mathbb{Z}$, satisfies the condition $\Sigma_{n=-\infty}^{\infty} |c_n| |\hat{f}(n)| < \infty$ for all integrable f, then $\Sigma_{n=-\infty}^{\infty} |c_n| < \infty$.

8.42 Prove the Paley theorem: The Fourier series of a function in $C(\mathbb{T})$ with nonnegative Fourier coefficients converges absolutely and uniformly on $[0, 2\pi]$.

8.43 Prove that the Fourier series of a continuous even function on $[-\pi, \pi]$ convex on $[0, \pi]$ converges absolutely.

8.44 Show that if f is a continuous even function on $[-\pi, \pi]$ and its Fourier coefficients a_k are nonnegative, then $\Sigma_{k=1}^{\infty} ka_k < \infty$ if and only if $\Sigma_{n=1}^{\infty} n \int_0^{\pi/n} (f(0) - f(x)) \, dx < \infty$.

8.45 Assume that the sequence (c_n), $n \in \mathbb{Z}$, is even, convex on $\mathbb{N}$ and decreases to zero if $|n| \to \infty$. Prove that the trigonometric series $\Sigma_{n=-\infty}^{\infty} c_n e^{inx}$ represents a positive integrable function.

8.46 Consider a trigonometric series $f(x) = \Sigma_{n=1}^{\infty} c_n \sin nx$, $c_n \geq c_{n+1}$, $\lim_{n\to\infty} c_n = 0$ (f is defined everywhere, see 8.2). Show that $f \in L^1$ if and only if $\Sigma_{n=1}^{\infty} c_n/n < \infty$. In this case $\Sigma_{n=1}^{\infty} c_n \sin nx$ is the Fourier series of f and $\lim_{n\to\infty} \|s_n f - f\|_1 = 0$.

8.47 Prove the following asymptotic formulas for the Fourier coefficients of the function $f(x) = x^{-a}$, $0 < x \leqslant 2\pi$, $0 < a < 1$

$$a_n \simeq \frac{n^{a-1}}{2\Gamma(a)\cos(\pi a/2)} \qquad b_n \simeq \frac{n^{a-1}}{2\Gamma(a)\sin(\pi a/2)}$$

Observe that if $p < 1/a$ then $f \in L^p$, if $q < 1/(1-a)$ then $\sum_{n=1}^{\infty}(|a_n|^q + |b_n|^q) = \infty$. Compare this with the well-known Hausdorff-Young theorem on the Fourier coefficients of L^p functions.

8.48 Consider the odd function f which is equal to $k^a \cos(k^2 x)$ in the interval $(\pi/(k+1)^b, \pi/k^b)$, $k = 1, 2, \ldots,$ $0 < a < b < 1$. Show that f is integrable and that its Fourier coefficients in the sine Fourier expansion satisfy the estimate $b_n = O(n^{(a-1)/2} \log n)$.

  This example may show that for $q > 2$ the convergence of $\sum |b_n|^q$ does not imply that $f \in L^p$ for some $p = p(q) > 1$.

8.49 Prove that the function

$$f(x) = -x + \lim_{n \to \infty} \int_0^x (1 + \cos t)(1 + \cos 4t) \cdots (1 + \cos 4^n t)\, dt$$

is continuous, 2π-periodic, of bounded variation, $\sum |\hat{f}(n)| < \infty$, but $n \int_{-\pi}^{\pi} f(x) \sin nx\, dx$ does not tend to zero and hence f is not absolutely continuous.

8.50 Show that $f(x) = \sum_{n=1}^{\infty} n^3(n^4 + 1)^{-1} \sin nx$ is infinitely differentiable in the open interval $(0, 2\pi)$, but it is discontinuous at $x = 0$.

8.51 The functions $f(x) = \sum_{n=1}^{\infty} n^{-1} \sin(x/2^n)$, $g(x) = \sum_{n=1}^{\infty} (-1)^n n^{-1} \sin(x/2^n)$ are uniformly continuous and $\mathbb{R}$-analytic. Is any of these functions bounded? What can be said about $h(x) = \sum_{n=1}^{\infty} (-1)^n n^{-1} \cos(x/2^n)$?

8.52 Show that for the function $f(x) = \operatorname{sgn} x$, $x \in (-\pi, \pi)$, there exists a constant $c > 1$ such that $\lim_{n \to \infty} \sup_{0 < x < \varepsilon} |s_n f(x)| > c$ for every $\varepsilon > 0$. It is one of the appearances of the Gibbs phenomenon. Prove that this appears in a neighborhood of any point of discontinuity of a piecewise smooth function.

8.53 An other interpretation of the Gibbs phenomenon: Let f be piecewise smooth function on $(-\pi, \pi)$ and $G_n \in (-\pi, \pi) \times \mathbb{R}$ be the graph of the nth partial sum of its Fourier series. Study the set of cluster points of the sequence (G_n).

8.54 Find the Fourier series of the function $f(t) = t$ in $[0, 2\pi)$. Determine the minimal positive t_n such that the sum $s_n(t)$ of the first $n + 1$ terms of the Fourier series of f has a local minimum in t. Show that $\lim_{n \to \infty} s_n(t_n) = \pi - 2\int_0^{\pi} t^{-1} \sin t\, dt$. How can one explain that $f(t_n) - s_n(t_n)$ does not converge to zero if n tends to infinity?

8.55 Let S denote the class of functions represented as finite sums of terms like Ax^ke^{bx}, where $A, b \in \mathbb{C}$, $k \in \mathbb{N}$. Show that

(i) If $\sum_{n=1}^{\infty} r(n) \sin nx \in S$ and the function f is even and periodic on $\mathbb{Z}$, then $\sum_{n=1}^{\infty} r(n) f(n) \sin nx \in S$.

(ii) If $\sum_{n=1}^{\infty} r(n) \cos nx \in S$ and the function f is odd and periodic on $\mathbb{Z}$, then $\sum_{n=1}^{\infty} r(n) f(n) \cos nx \in S$.

8.56 Let (a_k), $k \in \mathbb{Z}$, be an odd periodic complex sequence with period n. Show that $a_k = \sum_{r=0}^{n-1} b_r e^{2\pi irk/n}$, where $b_k = n^{-1} \sum_{r=0}^{n-1} a_r e^{-2\pi irk/n}$ and $\sum_{k=1}^{\infty} a_k/k = -\pi in^{-1} \sum_{r=0}^{n-1} rb_r$.

8.57 For a strictly positive function $f(x) = a_0 + 2\sum_{n=1}^{\infty} (a_n \cos nx + b_n \sin nx)$ consider the roots $\lambda_{n1}, \ldots, \lambda_{nn}$ of the characteristic equation

$$\det \begin{bmatrix} a_0 - \lambda & a_1 - ib_1 & \cdots & a_{n-1} - ib_{n-1} \\ a_1 + ib_1 & a_0 - \lambda & \cdots & a_{n-2} - ib_{n-2} \\ \hdashline a_{n-1} + ib_{n-1} & a_{n-2} + ib_{n-2} & \cdots & a_0 - \lambda \end{bmatrix} = 0$$

Prove that $\lim_{n \to \infty} (\lambda_{n1} \cdots \lambda_{nn})^{1/n} = \exp((2\pi)^{-1} \int_0^{2\pi} \log f(x)\, dx)$.

8.58 Prove the Kolmogorov theorem: If $f \in L^2$ and (n_k) is a lacunary sequence, that is, $n_{k+1}/n_k > 1 + \delta$ for some $\delta > 0$, $k = 1, 2, \ldots$, then $\lim_{k \to \infty} s_{n_k} f(x) = f(x)$ a.e.

8.59 Prove that if the Fourier series of the integrable function f is lacunary, that is, $f(x) \simeq \sum_{k=1}^{\infty} (a_{n_k} \cos n_k x + b_{n_k} \sin n_k x)$ with a lacunary sequence (n_k), then it converges to f a.e.

8.60 Show that if $f = \sum_{n=0}^{\infty} a_n e^{inx}$ is in $L^1(0, 2\pi)$, then there exists a sequence (n_k) of positive integers such that $n_{k+1}/n_k \to 1$ and $\sum_{k=1}^{\infty} |a_{n_k}| < \infty$.

8.61 Prove the Steinhaus theorem: If g is a 2π-periodic Lipschitz function, then for every integrable function f the difference $s_n(gf) - gs_n(f)$ tends to zero uniformly.

8.62 Show that if f is a 2π-periodic measurable function and $T_a f = f$ a.e. for a translation T_a with a/π irrational, then $f = $ const a.e.

8.63 Prove that for every $f \in L^1(0, 2\pi)$ $\lim_{n \to \infty} \|f^{*n}\|_1^{1/n} = \max_{n \in \mathbb{Z}} |\hat{f}(n)|$, where $f^{*n} = f * \cdots * f$ is an n-fold convolution of f.

8.64 Show that the set $\{f \in L^2(0, 2\pi): \lim_{n \to \infty} \sum_{k=-n}^{n} \hat{f}(k) \text{ exists}\}$ is a dense first category subset in $L^2(0, 2\pi)$.

8.65 Find necessary and sufficient conditions on the function g that guarantee the existence of $f \in L^2(\mathbb{T})$ satisfying $f * f = g$.

8.66 Show that if (K_N) is a sequence of nonnegative integrable functions on $\mathbb{T}$ such that $\lim_{N \to \infty} \hat{K}_N(n) = 1$ for all integers n, then $\lim_{N \to \infty} K_N * f = f$ uniformly for every continuous function f.

8.67 Consider a generalized convolution $*_s$ for a complex sequences indexed by $\mathbb{N}$ determined by a kernel $s : \mathbb{N} \times \mathbb{N} \to \mathbb{C} \backslash \{0\} : (f *_s g)(n) = \Sigma_{j+k=n} s(j, k) f(j) g(k)$. Show that if $*_s$ is associative, then there exists a function $t : \mathbb{N} \to \mathbb{C} \backslash \{0\}$ such that $f *_s g = t^{-1}(tf * tg)$, where $*$ is the usual convolution with $s \equiv 1$. The $*_s$ convolution is then commutative and s is a symmetric function.

8.68 For $n = 1, 2, \ldots$ define the functions

$$\omega_n(t) = \overline{\omega_{-n}(t)} = \left(\frac{|n|}{\pi}\right)^{1/2} \prod_{k=1}^{n-1} \frac{t - ki}{t + ki} \frac{1}{t + ni}$$

Show that (ω_n) is an orthonormal system in $L^2(\mathbb{R})$, that is, $\int_{-\infty}^{\infty} \omega_n(t) \overline{\omega_m(t)} \, dt = \delta_{nm}$. Moreover check that ω_n are the eigenfunctions of the Hilbert transform

$$\text{p.v.} \left(\frac{1}{\pi} \int_{-\infty}^{\infty} \omega_n(t) \frac{dt}{t - x}\right) = -(i \, \text{sgn} \, n) \omega_n(x)$$

8.69 The system (f_n) of measurable bounded functions defined on the interval $[a, b]$ is called maximal if there exists a set N of measure zero such that for every $x \neq y \in [a, b] \backslash N$ there exists an index n satisfying $f_n(x) \neq f_n(y)$. Prove the Rényi theorem: If the system (f_n) is maximal on $[a, b]$, then the system of products $(f_1^{m_1} f_2^{m_2} \cdots f_n^{m_n})$, $m_k = 0, 1, \ldots, k = 1, \ldots, n, n = 1, 2, \ldots$, is complete in $L^2(a, b)$.

Corollary: since $(\cos x, \sin x)$ is a maximal system on $[0, 2\pi]$ one obtains a new proof of the classical Riesz-Fischer theorem.

8.70 Show that the Gram-Schmidt orthogonalization procedure applied to the system of monomials (x^n) in the space $L^2((a, b), d\mu(x))$ gives the system of orthogonal polynomials $(p_k(x))$ satisfying the recurrent relation

$$a_k a_{k+1}^{-1} p_{k+1}(x) = (x - b_k) p_k(x) - a_{k-1} a_k^{-1} p_{k-1}(x)$$

where a_k is the leading coefficient of p_k and b_k is a real number.

8.71 Prove that in the notation of the preceding problem the Christoffel-Darboux formula holds:

$$\sum_{k=0}^{n} p_k(x) p_k(t) = a_n a_{n+1}^{-1} (p_n(x) p_{n+1}(t) - p_n(t) p_{n+1}(x)) / (t - x)$$

8.72 Assume that a function $f \in L^2((a, b), w(x) \, dx)$ satisfies at $x_0 \in (a, b)$ the Dini-Lipschitz condition with exponent $a > 1$: $|f(x_0 + h) - f(x_0)| \leqslant$

$C|\log|h||^{-a}$ for sufficiently small $|h|$, $(p_k(x))$ is the orthonormal system of polynomials and $(p_k(x))$ and $w(x)$ are bounded in a neighborhood of x_0.

Prove that the partial sums of the orthogonal expansion of f with respect to (p_k) at x_0 tend to $f(x_0)$.

8.73 Prove the Favard theorem: If the system $(p_k(x))$ of polynomials with leading coefficients equal to 1 satisfies the recurrent relations $p_{k+1}(x) = (x - A_k)p_k(x) - B_k p_{k-1}(x)$, for some A_k, B_k, $B_k > 0$, then there exists a measure μ such that $(p_k(x))$ is the orthogonal system in $L^2((a, b), d\mu(x))$.

8.74 Prove that the Haar series of the function $f \in L^1(0, 1)$ converges to f a.e. Moreover this series converges to the value of f at the points of continuity of f (and the convergence is uniform on the intervals of uniform continuity of f).

Recall that the Haar system on $[0, 1]$ contains the functions $\chi_0^{(0)} \equiv 1$, $\chi_0^{(1)} = 1$ on $[0, 1/2]$ and -1 on $(1/2, 1]$, $\chi_m^{(k)} = 2^{m/2}$ on $((k-1)/2^m, (k-1/2)/2^m)$, $= -2^{m/2}$ on $((k-1/2)/2^m, k/2^m)$ and $= 0$ otherwise, $k = 1, \ldots, 2^m$, $m = 1, 2, \ldots$.

8.75 Let $r_1(x) = \chi_0^{(1)}(x)$, $r_{n+1}(x) = 2^{-n/2} \sum_{k=1}^{2^n} \chi_n^{(k)}(x)$, where $\chi_m^{(k)}$ are the Haar functions from the preceding problem, $n = 1, |2, \ldots$, $0 < x < 1$, $r_n(0) = r_n(1) = 0$. Observe that one obtains in this way the Rademacher system $r_n(x) = \operatorname{sgn} \sin(2^n \pi x)$. Show that $(r_n(x))$ is not complete in $L^2(0, 1)$. Prove that $\sum_{n=1}^{\infty} c_n r_n(x)$ converges a.e. if and only if $\sum_{n=1}^{\infty} c_n^2 < \infty$.

8.76 Show the completeness of the Walsh system $(w_n(x))$ on $[0, 1]$, where $w_0 \equiv 1$, $w_n = r_{k_1+1} r_{k_2+1} \cdots r_{k_p+1}$ if $n = 2^{k_1} + \cdots + 2^{k_p}$, $k_1 < \cdots < k_p$, r_k are the Rademacher functions. Express the Haar function using the Walsh ones.

8.77 Prove the Menšov-Rademacher theorem: If $\sum_{n=1}^{\infty} c_n^2 (\log n)^2 < \infty$ and $(f_n(x))$ is any orthonormal system in $L^2((a, b), d\mu(x))$, then the series $\sum_{n=1}^{\infty} c_n f_n(x)$ converges a.e.

8.78 Deduce the following corollary from the Menšov-Rademacher theorem: If $\sum_{n=1}^{\infty} c_n^2 < \infty$ then for the partial sums of the orthogonal series $\sum_{n=1}^{\infty} c_n f_n(x)$ the estimate $s_n(x) = o(\log n)$ holds a.e.

8.79 Prove the Talalian theorem: Given an orthonormal system (f_n) in $L^2(a, b)$ there exists a series $\sum_{n=1}^{\infty} c_n f_n(x)$, $c_n \to 0$, such that for every measurable function there exists a subsequence of partial sums converging to this function a.e.

8.80 Suppose that $1 = t_0 > t_1 > \cdots > t_k > 0$ and $\lim_{n \to \infty} \sum_{j=0}^{k} a_j \cos(n t_j \pi) = 0$. Does it imply that $a_j = 0$ for all j?

8.81 Are the functions $\exp(i \cos(t - t_j))$ $j = 1, \ldots, k$, $0 \leqslant t_1 < t_2 < \cdots < t_k < 2\pi$, linearly independent over the complex numbers?

8.82   Determine when the function $s(t) = \sum_{j=1}^{N} a_j \exp(i\lambda_j t)$, $\lambda_j \in \mathbb{R}$, has a limit for $t \to \infty$.

8.83   Show that if $f \in L^1(\mathbb{R})$, $f \geqslant 0$, $\hat{f} \geqslant 0$, then for every natural k $\int_{-k}^{k} f \leqslant (2k+1) \int_{-1}^{1} f$.

8.84   Show that the linear subspace spanned by the translates $(T_s f)$, $s \in \mathbb{T}$, of a function $f \in L^2(\mathbb{T})$, is dense if and only if $\hat{f}(n) \neq 0$ for all $n \in \mathbb{Z}$.

8.85   Show that the linear subspace spanned by the translates of $\chi_{[0,a]} + \chi_{[0,b]}$ is dense in $L^1(\mathbb{R})$ if and only if a/b is irrational.

8.86   Show that if $f \in L^1(\mathbb{R})$, $f \geqslant 0$ and f vanishes outside a finite interval where f strictly decreases, then the linear subspace spanned by the translates of f is dense in $L^1(\mathbb{R})$.

8.87   Determine the spectrum of the Fourier transform in $L^2(\mathbb{R}^n)$.

8.88   Give an explicit formula for the operator $f(\mathscr{F})$, where f is a continuous function on the unit circle and $\mathscr{F}$ is the Fourier transform on $L^2(\mathbb{R})$.

8.89   Define $u_a(x, t) = (2\pi)^{-n/2} \int_{|s| \leqslant 1} \exp(-i(x_1 s_1 + \cdots + x_n s_n)) g_a(s_1, \ldots, s_n, t) \, ds_1 \ldots ds_n$, where $g_a(s_1, \ldots, s_n, t) = |s|^{-a-2}(1 - \exp(-|s|^2 t))$, $|s| = (s_1^2 + \cdots + s_n^2)^{1/2}$, $t \geqslant 0$, $a \in \mathbb{R}$. Find a such that $\lim_{t \to \infty} \int_{\mathbb{R}^n} |u_a(x, t)|^2 \, dx_1 \cdots dx_n = \infty$.

8.90   Prove the Poisson summation formula: $\sum_{n=-\infty}^{\infty} f(2\pi n) = (2\pi)^{-1} \sum_{n=-\infty}^{\infty} \hat{f}(n)$ for $f \in \mathscr{S}$ and more generally if f and $\hat{f}$ are continuous and satisfy the estimates $|f(x)| \leqslant c(1 + |x|)^{-1-\delta}$, $|\hat{f}(\xi)| \leqslant c(1 + |\xi|)^{-1-\delta}$ for some $\delta > 0$. Apply this result to the series $\sum_{n=-\infty}^{\infty} (1 + n^2)^{-1}$, $\sum_{n=1}^{\infty} n^{-2}$.

8.91   Given two complex sequences (a_k) and $(b_k) \in \ell^1(\mathbb{Z})$ such that $\sum_{n=-\infty}^{\infty} a_n = \sum_{n=-\infty}^{\infty} b_n$ does there exist a continuous function in $L^2(\mathbb{R})$ satisfying $f(n) = a_n$ and $\tilde{f}(n) = \int e^{2\pi i n x} f(x) \, dx = b_n$? Is f determined uniquely?

8.92   Show that the Jacobi theta function $\theta(x) = \sum_{n=-\infty}^{\infty} \exp(-\pi n^2 x)$ is smooth on $\mathbb{R}^+$ and satisfies the identity $\theta(x) = x^{-1/2} \theta(1/x)$.

8.93   Prove the Kotelnikov-Shannon sampling theorem (the cardinal interpolation series): If f is the Fourier transform of a function $g \in L^2(\mathbb{R})$ vanishing outside the interval $[-a, a]$, then f may be represented as $f(x) = \lim_{k \to \infty} \sum_{n=-k}^{k} f(n\pi/a) \sin a(x - n\pi/a)/(a(x - n\pi/a))$. The convergence here is uniform on $\mathbb{R}$ and in $L^2(\mathbb{R})$.

8.94 P is a polynomial on $\mathbb{R}$ of degree $2m$, $m \geqslant 1$, without real roots. Show that the Fourier transform of $1/P(x)$ is smooth outside the origin, where all one-sided derivatives exist. How many continuous derivatives has it?

8.95 Show that $\int_0^\infty t^{-1}e^{-t}\sin st\,dt = \arctan s$. Compute the Fourier transform of the function $k(t) = 2^{-1}\int_{|t|}^\infty x^{-1}e^{-x}\,dx$. Show that $1 - (2\pi)^{1/2}\hat{k}(s)$ has only the double zero at $s = 0$ in the strip $|\text{Im } s| < 1$. Study the solutions of the integral equation $f(x) = \int_{-\infty}^\infty k(x - t)f(t)\,dt$ satisfying the estimate $f(x) = O(e^{|x|})$ for $|x| \to \infty$.

8.96 The function f is positive and integrable and strictly decreases on $\mathbb{R}^+$, $\text{Re }\hat{f}$ also strictly decreases on $\mathbb{R}^+$. Show that $F(z) = \int_0^\infty f(t)e^{izt}\,dt$ is analytic and univalent in the half-plane $\text{Im } z > 0$.

8.97 Prove that if $f \in L^1(\mathbb{R})$, f and f' are absolutely continuous, and f' and $f'' \in L^1(\mathbb{R})$, then the Fourier transform of f is in $L^1(\mathbb{R})$. The assumptions above are satisfied for example if $f, f', f'' \in L^1(\mathbb{R}) \cap C_0(\mathbb{R})$.

8.98 Prove that if μ is a finite positive measure, then $F(t) = \int_1^\infty \cos tx\,d\mu(x)$ has a zero in the interval $[0, \pi]$ and π cannot be replaced by any smaller number.

8.99 Show that the characteristic function of the Cantor measure v $F(t) = \int e^{ixt}\,dv(x)$ can be represented in the form $F(t) = e^{it/2}\prod_{n=1}^\infty \cos(3^{-n}t)$. Show that $F(t)$ does not converge to zero if t tends to infinity. Moreover, if A is the operator of multiplication by x in $L^2((0, 1), dv(x))$, then $\exp(itA)$ does not weakly converge to zero if t tends to infinity.

8.100 (a_n), $n \in \mathbb{Z}$, is a complex sequence such that there exists $K > 0$ satisfying for all $N \in \mathbb{N}$ and all complex sequences (c_n), $n \in \mathbb{Z}$, the inequality $|\Sigma_{n=-N}^N a_n c_n| \leqslant K \sup_t |\Sigma_{n=-N}^N c_n e^{int}|$. Prove that there exists a complex Borel measure μ on $[0, 2\pi]$ such that $|\mu| \leqslant K$ and $a_n = \int_0^{2\pi} e^{-int}\,d\mu(t)$.

8.101 Define for a complex measure m on $\mathbb{T}^2$ the function $f(K, N) = (2K + 1)^{-1}(2N + 1)^{-1}\Sigma_{k=-K}^K \Sigma_{n=-N}^N \hat{m}(k, n)$. Show that $\lim_{K \to \infty}\lim_{N \to \infty} f(K, N) = \lim_{N \to \infty}\lim_{K \to \infty} f(K, N)$.
    Does there exist a complex Borel measure such that $\hat{m}(k, n) = k^2/(1 + k^2 + n^2)$?

8.102 Prove the Wiener theorem: If m is a finite Borel measure on $\mathbb{R}$, $F(t) = \int_{-\infty}^\infty e^{ixt}\,dm(x)$ its Fourier transform, then $\lim_{T \to \infty}(2T)^{-1}\int_{-T}^T |F(t)|^2\,dt = \Sigma_{x\in\mathbb{R}} |m(\{x\})|^2$ and $\lim_{N \to \infty}(2N + 1)^{-1}\Sigma_{n=-N}^N |F(n)|^2 = \Sigma_{x\in\mathbb{R}} |m(\{x\})|^2$.
    Prove the analogous result for measures on $\mathbb{T}$.
    How can one generalize this result in the situation where, instead of the Fourier transform $F = \mathscr{F}_m$ one considers a unitary operator in a separable Hilbert space with eigenvalues (λ_j) and the spectral projections (P_j) corresponding to (λ_j)?

8.103 Let (w_n), $n \in \mathbb{Z}$, be a sequence of positive numbers such that $w_0 = 1$, $w_{n+m} \leqslant w_n w_m$. Show that $\lim_{n \to \infty} w_n^{1/n} = R_+ \geqslant \lim_{n \to \infty} w_{-n}^{1/n} = R_-$.

Define $\mathbb{A}$ as the set of all complex sequences on $\mathbb{Z}$ satisfying the condition $\|f\| = \Sigma_{n=-\infty}^{\infty} |f(n)|w_n < \infty$. $\mathbb{A}$ with the convolution $(f * g)(n) = \Sigma_{k=-\infty}^{\infty} f(n-k)g(k)$ is a commutative Banach algebra. Identifying the space of maximal ideals of $\mathbb{A}$ with the annulus $\{z: R_- \leqslant |z| \leqslant R_+\}$ describe the Gelfand transform in $\mathbb{A}$.

Consider the special cases:

$$w_n \equiv 1$$

$$w_n = 2^n \qquad w_n = \begin{cases} 1 & \text{for } n < 0 \\ 2^n & \text{for } n \geqslant 0 \end{cases}$$

$$w_n = 1 + 2n^2 \qquad w_n = \begin{cases} 1 & \text{for } n < 0 \\ 1 + 2n^2 & \text{for } n \geqslant 0 \end{cases}$$

Determine when $\mathbb{A}$ is selfadjoint, that is, $\bar{\mathbb{A}} = \mathbb{A}$, and when $\mathbb{A}$ is semisimple.

Does there exist a sequence (w_n) such that the space of maximal ideals of $\mathbb{A}$ is a circle and $\hat{\mathbb{A}}$ contains smooth functions only?

8.104 Prove the following version of the Heisenberg-Weyl inequality: For every $p \in (1, 2]$ there exists a constant $C(p)$ such that for $f \in \mathscr{S}(\mathbb{R})$, $f(0) = 0$, the inequality $\|f\|_2^2 \leqslant C(p)\|xf\|_p \|\xi\hat{f}\|_p$ holds.

9

Functional Analysis

Topics covered include:
Matrices and determinants, Hilbert and Banach spaces, linear operators, and distributions.

9.1 Consider a matrix (a_{ij}), $i, j = 1, \ldots, n$, such that $a_{ij} \in \{0, 1, \ldots, 9\}$. Show that if a prime number p divides $\sum_{j=1}^{n} a_{ij} 10^{n-j}$ for $i = 1, \ldots, n$, then p divides $\det(a_{ij})$.

9.2 $\{e_1, \ldots, e_n\}$ is a basis in $\mathbb{R}^n$. Show that there exists a constant $k > 0$ such that every set of vectors $\{x_1, \ldots, x_n\}$ satisfying the inequalities $|e_j - x_j| < k$, $j = 1, \ldots, n$, is also a basis in $\mathbb{R}^n$.

9.3 A is a linear operator in $\mathbb{R}^n$, $c_1, \ldots, c_n$ are the columns of a $n \times n$ matrix. Calculate $\det(Ac_1, \ c_2, \ldots, c_n) + \det(c_1, \ Ac_2, \ldots, c_n) + \cdots + \det(c_1, c_2, \ldots, Ac_n)$.

9.4 Show that $\det(1/(i + j + 1))$, $i, j = 0, \ldots, n$, is positive for all n.

9.5 Show that if $F \geqslant 0$ is a decreasing function on $[0, 1]$ and $a_{ij} = \int_0^1 x^{j-1}(F(x))^{n-i+1} \, dx$, $i, j = 1, \ldots, n$, then $\det(a_{ij}) \geqslant 0$.

9.6 Calculate the generalized Vandermonde determinant $\det(a_{ij})$,

$i, j = 1, \ldots, n,$ where $a_{ij} = \binom{i-1}{k-1} x_q^{i-k}$ if $i \geqslant k$ and $a_{ij} = 0$ otherwise, $q = \min\{p : j \leqslant n_1 + \cdots + n_p\}, k = j - (n_1 + \cdots + n_{q-1}), n_i$ are given positive integers, $n_1 + \cdots + n_r = n.$

The classical Vandermonde determinant corresponds to $n_q = 1.$

9.7 Show that the determinants

$$D_n = \det \begin{bmatrix} f' & f & 0 & 0 & \cdots & 0 & 0 \\ f''/2! & f' & f & 0 & \cdots & 0 & 0 \\ f'''/3! & f''/2! & f' & f & \cdots & 0 & 0 \\ \hdashline f^{(n)}/n! & f^{(n-1)}/(n-1)! & \cdots & \cdots & \cdots & f''/2! & f' \end{bmatrix}$$

satisfy the recurrent relation $D_{n+1} = f'D_n - fD_n'/(n+1).$

9.8 Evaluate $\det(a_{ij}), i, j = 1, \ldots, n,$ where $a_{ij} = \sin^2(jt_i), t_1, \ldots, t_n \in \mathbb{R}.$

9.9 Show that for symmetric positive definite matrices A, B, C the real parts of the roots of the equation $\det(\lambda^2 A + \lambda B + C) = 0$ are negative.

9.10 Prove that if the matrix $F(A)$ is positive definite for every matrix A of the form $(a_{i-j}),$ then $F(z) = \sum_{n=0}^{\infty} c_n z^n$ and $c_n \geqslant 0.$

9.11 Let $B = (b_{ij}), i, j = 1, \ldots, n,$ where $b_{ij} = a_i \bar{a}_j / (1 - a_i \bar{a}_j)$ for different complex numbers $a_1, \ldots, a_n, 0 < |a_k| < 1.$ Show that B is positive definite and

$$\max_{z_k \in \mathbb{C}} \left\{ |z_1 + \cdots + z_n|^2 : \sum_{i,j=1}^{n} b_{ij} z_i \bar{z}_j = 1 \right\} = \prod_{k=1}^{n} |a_k|^{-2} - 1$$

9.12 Does the positive definiteness of the quadratic form $\sum_{i,j=1}^{n} a_{ij} x_i x_j$ imply that the form $\sum_{i=1}^{n} a_{ii} x_i^2 + \sum_{i \neq j} |a_{ij}| x_i x_j$ is positive definite?

9.13 Prove that if A, B are $n \times n$ symmetric real matrices and A is positive definite, then AB has n independent eigenvectors.

9.14 Characterize the closure of the set of diagonalizable matrices in the space of all $n \times n$ complex matrices.

9.15 Show that if the commutator of matrices A, B commutes with them, then it is nilpotent.

9.16 Let A be a matrix such that for all integers $k > 0 \operatorname{Tr}(A^k) = 0.$ Show that A is nilpotent.

9.17 Prove the formula

$$\det(I + \lambda A) = \exp\left(\sum_{k=1}^{\infty} (-1)^{k-1} \lambda^k \operatorname{Tr}(A^k)/k \right)$$

9.18   The polynomial $p(x) = x^n + a_1 x^{n-1} + \cdots + a_n$ is irreducible. Show that each nonzero matrix commuting with

$$
\begin{bmatrix}
0 & 1 & 0 & \cdots & 0 & 0 \\
0 & 0 & 1 & \cdots & 0 & 0 \\
0 & 0 & 0 & \cdots & 0 & 1 \\
-a_n & -a_{n-1} & -a_{n-2} & \cdots & -a_2 & -a_1
\end{bmatrix}.
$$

is invertible.

9.19   A linear operator A on $L = \mathbb{R}^n$ or $L = \mathbb{C}^n$ is said to be regular if A has a cyclic vector, that is, such a vector v that any A invariant linear subspace of L containing v is equal to L. Show that the following conditions are equivalent:
   (i) A is regular.
   (ii) The operators $I, A, A^2, \ldots, A^{n-1}$ are linearly independent.
   (iii) There exists a vector x such that $\{x. Ax, A^2x, \ldots, A^{n-1}x\}$ is a basis in L.
   (iv) The minimal polynomial of the matrix associated to A is its characteristic polynomial.
Prove that any matrix of the form

$$
\begin{bmatrix}
0 & 1 & 0 & \cdots & 0 \\
0 & 0 & 1 & \cdots & 0 \\
0 & 0 & 0 & \cdots & 1 \\
a_1 & a_2 & a_3 & \cdots & a_n
\end{bmatrix}
$$

defines a regular operator and, conversely, for any regular operator there exists a basis in which its matrix is of the above form.
   Show that two matrices of this form are similar if and only if they coincide.

9.20   Prove that each family of commuting unitary operators in $\mathbb{C}^n$ has a complete orthonormal system of eigenvectors.

9.21   Show that if λ is a simple eigenvalue of a matrix A, then the coordinates of the corresponding eigenvectors are polynomial functions of λ and the entries of A.

9.22   An analytic description of the Jordan normal form: A is an $n \times n$ matrix with eigenvalues $\lambda_1, \ldots, \lambda_s$. Define for $j = 1, \ldots, s$ the operators

$$
P_j = \frac{-1}{2\pi i} \oint_{|\lambda - \lambda_j| = \varepsilon_j} (A - \lambda)^{-1}\, d\lambda, \quad N_j = \frac{-1}{2\pi i} \oint_{|\lambda - \lambda_j| = \varepsilon_j} (\lambda - \lambda_j)(A - \lambda)^{-1}\, d\lambda
$$

for sufficiently small $\varepsilon_j > 0$.
   Show   that $P_j^2 = P_j$, $N_j P_j = P_j N_j = N_j$, $A P_j = P_j A = \lambda_j P_j + N_j$, $P_j P_k = 0$ for $j \neq k$, $\Sigma_{j=1}^s P_j = I$ and finally $A = \Sigma_{j=1}^s (\lambda_j P_j + N_j)$.

9.23 Solve the Pólya functional equation $Y(t + s) = Y(t)Y(s)$, $s, t \in \mathbb{R}$, where $Y(\cdot)$ is a continuous function with values in the space of nonsingular matrices.

9.24 It is easy to see that $\exp(tA)$, where A is an $n \times n$ matrix, has the form $\exp(tA) = u_0(t)I + u_1(t)A + \cdots + u_{n-1}(t)A^{n-1}$. Determine the differential equations satisfied by the coefficients $u_j(t)$.

9.25 Prove that the matrix exponent $\exp(tA)$ has positive entries for all $t \geqslant 0$ if and only if the entries of A satisfy the condition $a_{ij} \geqslant 0$ for $i \neq j$.

9.26 Prove the Schur inequality for the eigenvalues λ_j of the matrix (a_{ij}), $i, j = 1, \ldots, n$:

$$\sum_{j=1}^{n} |\lambda_j|^2 \leqslant \sum_{i,j=1}^{n} |a_{ij}|^2$$

9.27 Prove the Bendixson theorem: If λ is an eigenvalue of the real matrix (a_{ij}), then

$$|\operatorname{Im} \lambda| \leqslant \left(\frac{n(n-1)}{2}\right)^{1/2} \max_{i,j} |a_{ij} - a_{ji}|/2$$

Similarly for the complex matrix

$$|\lambda| \leqslant n \max_{i,j} |a_{ij}|, \quad |\operatorname{Re} \lambda| \leqslant n \max_{i,j} |a_{ij} + \overline{a_{ji}}|/2,$$

$$|\operatorname{Im} \lambda| \leqslant n \max_{i,j} |a_{ij} - \overline{a_{ji}}|/2$$

9.28 Show the Hadamard inequality

$$|\det C|^2 \leqslant \prod_{i=1}^{n} \sum_{j=1}^{n} |c_{ij}|^2$$

for every $n \times n$ complex matrix $C = (c_{ij})$.

9.29 Define for $p \neq 0$ and an $n \times n$ positive definite matrix C with eigenvalues $\lambda_1 \leqslant \cdots \leqslant \lambda_n$, $j_p(C) = (\sum_{k=1}^{n} \lambda_k^p/n)^{1/p}$. Observe that the definition of $j_p(C)$ can be extended by continuity for $p = 0, +\infty, -\infty$: $j_0(C) = (\det C)^{1/n}$, $j_{+\infty}(C) = \lambda_n$, $j_{-\infty}(C) = \lambda_1$.

Let $\Delta(C)$ denote the diagonal matrix with the same diagonal as C. Prove that $j_p(C) \leqslant j_p(\Delta(C))$ for $-\infty \leqslant p \leqslant 1$ and $j_p(C) \geqslant j_p(\Delta(C))$ for $p \geqslant 1 \geqslant +\infty$.

9.30 The matrix (a_{ij}), $i, j = 1, \ldots, n$, satisfies the following conditions: $\sum_{i=1}^{n} |a_{ij}| \leqslant 1$ for all j and $\sum_{j=1}^{n} |a_{ij}| \leqslant 1$ for all i. Show that for $1 \leqslant p, q \leqslant \infty$, $1/p + 1/q = 1$

$$\left|\sum_{i,j} a_{ij} x_i y_j\right| \leqslant \left(\sum_i |x_i|^p\right)^{1/p} \left(\sum_j |y_j|^q\right)^{1/q}$$

Deduce the inequality

$$|\mathrm{Tr}(AB)| \le (\mathrm{Tr}\,|A|^p)^{1/p}(\mathrm{Tr}\,|B|^q)^{1/q}$$

9.31 A, B are the complex matrices such that A is positive definite and $\|AB\| \le 1$, $\|BA\| \le 1$. Show that $\|A^{1/2}BA^{1/2}\| \le 1$.

9.32 Does there exist a constant k such that $|A + B| \le k(|A| + |B|)$ for all $n \times n$ matrices A and B? Here $|A| = (A^*A)^{1/2}$ and $C \le D$ means: $D - C$ is positive definite.

9.33 Prove that if matrices A and B are strictly positive definite then for $0 \le t \le 1$ $(tA + (1 - t)B)^{-1} \le tA^{-1} + (1 - t)B^{-1}$.

9.34 Show that for positive definite matrices A and B and $0 \le t \le 1$, $\det(tA + (1 - t)B) \ge (\det A)^t(\det B)^{1-t}$.

9.35 Prove the Birkhoff theorem: Every $n \times n$ stochastic matrix (a_{ij}), that is, such that $a_{ij} \ge 0$, $\Sigma_{i=1}^{n} a_{ij} = \Sigma_{j=1}^{n} a_{ij} = 1$, $i, j = 1, \ldots, n$, can be represented as a convex combination of (at most $n!$) permutation matrices, that is, the matrices of linear operators changing the coordinates with respect to the basis.

9.36 Show that for a given $n \times n$ matrix A the matrix equation $AX - XA = cX$ has nontrivial solutions if and only if $c = \lambda_i - \lambda_j$, where $\lambda_1, \ldots, \lambda_n$ are the eigenvalues of A.

9.37 Show that for the matrix

$$A = \begin{bmatrix} p & q & r \\ r & p & q \\ q & r & p \end{bmatrix}, \text{ where } p, q, r > 0, \ p+q+r=1, \ \lim_{n \to \infty} A^n = \frac{1}{3}\begin{bmatrix} 1 & 1 & 1 \\ 1 & 1 & 1 \\ 1 & 1 & 1 \end{bmatrix}.$$

9.38 Determine all matrices B such that $B = \lim_{n \to \infty} t_n A^n$ for a sequence of reals (t_n) and a matrix A.

9.39 Determine the conditions on the matrices A and B guaranteeing that the recurrent matrix sequence $X_{n+1} = X_n(2I - AX_n)$, $X_0 = B$, converges. Find the limit of X_n.

9.40 $A(t)$ is a continuous matrix valued function on $\mathbb{R}^+$ such that $A(t) + A(t)^* \ge 0$ and the function $X(t) = \mathrm{Re}(\mathrm{Tr}\,A(t))$ satisfies the condition $\limsup_{t \to \infty} \int_0^t X(s)\,ds < \infty$.
Show that all the solutions of the system of linear differential equations $\dot{y} = Ay$ are bounded.

9.41 Prove that each continuous function f defined for $n \times n$ complex matrices such that $f(BAB^{-1}) = f(A)$ for all A and invertible B is of the form

$f(A) = F(c_1(A), \ldots, c_n(A))$, where the invariants c_i are defined by the identity $\det(\lambda I - A) = \lambda^n - c_1(A)\lambda^{n-1} + \cdots + (-1)^n c_n(A)$.

9.42 Prove the functional analog of the Sylvester theorem: If $U \subseteq \mathbb{R}^k$ is a contractible region with smooth boundary and A is a smooth mapping on U with values in $n \times n$ symmetric matrices, then the following conditions are equivalent:
  (i) There exists a smooth function $G: U \to GL(n, \mathbb{R})$ such that GAG^T is a diagonal matrix with p entries equal to 1 and q equal to -1, $p + q \leqslant n$.
  (ii) A has exactly p positive and q negative eigenvalues at each point of U.

9.43 Is the set $\{f \in C^\infty(0, 1): f(0) = 0, \int_0^1 f(x)x^{-1}\, dx = 0\}$ dense in $L^2(0, 1)$?

9.44 Let x, y, z be the points in a complex Hilbert space, $L(y, z)$ denotes the line containing y, z, $a = \|x - y\|$, $b = \|y - z\|$, $c = \|z - x\|$, $s = (a + b + c)/2$. Compare two generalizations of the definition of area of the triangle x, y, z:

$$A_1 = \frac{b}{2} d(x, L(y, z)) \qquad A_2 = \sqrt{s(s - a)(s - b)(s - c)}$$

9.45 Consider the functions $e_n(t) = e^{int}$, $n \in \mathbb{Z}$, $f_n = e_{-n} + ne_n$, $n = 1, 2, \ldots$, in $L^2(-\pi, \pi)$. Let $X_1 = \overline{\mathrm{lin}(e_0, e_1, e_2, \ldots)}$, $X_2 = \overline{\mathrm{lin}(f_1, f_2, \ldots)}$. Show that $X_1 \cap X_2 = \{0\}$, $X_1 + X_2$ is dense in $L^2(-\pi, \pi)$ but not closed. Moreover the natural projection from $X_1 + X_2$ onto X_1 (with kernel equal to X_2) is not continuous.

9.46 Give an example of three closed subspaces X_1, X_2, X_3 of an infinite dimensional Hilbert space such that $X_1 \supseteq X_3$, $\overline{\mathrm{lin}((X_1 \cap X_2) \cup X_3)} \not\subseteq X_1 \cap \overline{\mathrm{lin}(X_2 \cup X_3)}$.

9.47 Let $\{x_n : n \in \mathbb{N}\}$ be an orthonormal system in a Hilbert space, $E = \{x_m + mx_n : n > m, m, n \in \mathbb{N}\}$. Show that the closure of E in the weak topology contains 0 but there are no sequences weakly convergent to 0 in E.

9.48 Prove that any positive functional h on $L^2(M, \mu)$, where M is a measure space with a σ-finite nonatomic measure μ, cannot majorize the norm, that is, $\forall h \exists f \geqslant 0\ h(f) < \|f\|$.

9.49 Consider the subset $C = \{f : |f(t)| \leqslant 1 \text{ a.e.}\} \subseteq L^2((0, 1), \mathbb{R}^k)$. Show that for every $f \in C = \mathrm{bd}\ C$ there exists a continuous functional h on $L^2((0, 1), \mathbb{R}^k)$ such that $h(f) = \sup_{g \in C} h(g)$ if and only if the set $\{f : |f(t)| = 1\}$ is of positive Lebesgue measure. Is this supporting functional h determined uniquely?

9.50 X is a linear subspace of $L^2(0, 1)$ such that $|f(x)| \leqslant C\|f\|$ a.e. for a constant $C > 0$ and each $f \in X$. Show that $\dim X \leqslant C^2$.

9.51 Is it true that a closed subspace of $L^2(0, 1)$ contained in $L^p(0, 1)$ for some $p > 2$ must be finite dimensional?

9.52 Define for $0 < a \leqslant 1, 0 < x < 1$, the functions $f_a(x) = [a/x] - a[1/x]$. Is $\lin\{f_a : 0 < a \leqslant 1\}$ dense in $L^2(0, 1)$?

9.53 Consider the monomials $f_n(x) = x^n$, $n = 0, 1, 2, \ldots$, in the Hilbert space $H = L^2(\mathbb{R}, \exp(-x^2)\, dx)$. Show that $\sum_{n=0}^N f_n(ik)^n/n!$ converges to $\exp(ikx)$ in H. Prove that if $v \in H$ is orthogonal to all f_n, then $v = 0$.

Use that Gram-Schmidt procedure to obtain the basis of H composed of the Hermite polynomials.

9.54 It is well-known that for an orthonormal sequence $(x_n) \subseteq L^2(0, 1)$ the following conditions are equivalent:
  (i) For every $f \in L^2(0, 1)$ $\sum_{n=1}^N (f, x_n)x_n \to f$ as $N \to \infty$.
  (ii) $\bigcup_{n \in \mathbb{N}} \lin\{x_1, \ldots, x_n\}$ is dense in $L^2(0, 1)$.
  (iii) For every $f \in L^2(0, 1)$ $\sum_{n=1}^\infty |(f, x_n)|^2 = \|f\|^2$.
  (iv) (x_n) is the maximal orthonormal system.
  (v) $(f, x_n) = 0$ for all n implies $f = 0$.
  (vi) f is determined in the unique way by its coefficients (f, x_n).
Show that for the Riemann square integrable functions $f \in R^2(0, 1)$ the conditions (i), (ii), and (iii) are equivalent, (iv), (v), and (vi) are equivalent, (iii) implies (iv), but (iv) does not imply (iii).

9.55 Show that if (e_n) is an orthonormal basis in a separable Hilbert space and (f_n) is an orthonormal system such that $\sum_{n=1}^\infty \|e_n - f_n\| < \infty$, then (f_n) is also a basis.

9.56 Prove the Paley-Wiener theorem: Let (e_n) be an orthonormal basis in a separable Hilbert space H and let the sequence (f_n) satisfy the inequality $\|\sum a_n(e_n - f_n)\| \leqslant t^2 \sum |a_n|^2$ for some $0 < t < 1$ and every finite sequence of complex numbers (a_n). Then there exists a sequence $(h_n) \subseteq H$ such that $(f_m, h_n) = \delta_{mn}$ and for every $f \in H$ the expansions $f = \sum_{n=1}^\infty (f, h_n)f_n = \sum_{n=1}^\infty (f, f_n)h_n$ satisfy the inequalities

$$(1 + t)^{-1}\|f\| \leqslant \left(\sum_{n=1}^\infty |(f, h_n)|^2 \right)^{1/2} \leqslant (1 - t)^{-1}\|f\|$$

$$(1 - t)\|f\| \leqslant \left(\sum_{n=1}^\infty |(f, f_n)|^2 \right)^{1/2} \leqslant (1 + t)\|f\|.$$

9.57 Show that the functions $f_n(x) = (2\pi)^{-1/2} \exp(ic_n x)$, $n \in \mathbb{Z}$, $\max_n |c_n - n| < (\log 2)/\pi$, form a complete normalized system in $L^2(0, 2\pi)$.

9.58 Let (e_n) and (f_n), $n \in \mathbb{N}$, be two complete systems in $L^2(M, \mu)$, where μ is a σ-finite measure. Show that $(e_n(x)f_m(y))$, $(n, m) \in \mathbb{N} \times \mathbb{N}$, is a complete system in $L^2(M \times M, \mu \times \mu)$.

9.59 Prove that the functions $f_n(x) = \sin c_n x$, $n = 1, 2, \ldots$, where $c_1 < c_2 < \cdots$ are the positive roots of the equation $\tan c = c$ form a complete orthogonal system in $L^2(0, 1)$.

9.60 (f_n) is an orthonormal system in $L^2(0, 1)$ such that the sequence $g_n(x) = \Sigma_{k=0}^n f_k(x) \int_0^x f_k(t)\, dt$ is uniformly bounded and converges a.e. to a function f. Show that (f_n) is the complete system if and only if $f = 1/2$ a.e.

9.61 (f_n) is an orthonormal system in the Hilbert space $L^2(M, \mu)$, where μ is a σ-finite measure. Show that if $E = \{x : \lim_{n \to \infty} f_n(x) = f(x) \text{ exists}\}$, then $f = 0$ a.e. on E.

9.62 Prove that $\sup_n V(f_n, [0, 1]) = \infty$ for every orthonormal sequence (f_n) in $L^2(0, 1)$.

9.63 Consider the compact convex subset of ℓ^2 defined by $C = \{x = (x_n) \in \ell^2 : |x_n| \leqslant 3^{-n}\}$ and the functional $h : C \to \mathbb{C}$ such that $h(x) = \Sigma_{n=1}^\infty 2^n x_n$. Show that h is a homogeneous, additive and continuous functional on C with no extension to whole space ℓ^2.

9.64 Let P, Q be the orthogonal projections onto the subspaces X, Y of a Hilbert space. Determine the strong limit of $(PQ)^n$ as $n \to \infty$.

9.65 Show that a nontrivial convex combination of orthogonal projections in a Hilbert space cannot be a projection. Observe, by constructing an example, that the orthogonality assumption is important.

9.66 Let S be an isometry of the Hilbert space H. Show that there exists a closed subspace $X \subseteq H$ and an unitary operator U on H such that:

  (i) $X^\perp \subseteq \text{Ran } S$.
  (ii) $U_{|X} = S_{|X}$.
  (iii) $U_{|X^\perp} = S_{|X^\perp}^{-1}$.

9.67 Determine when the set of all invertible operators is dense (in norm topology) in the space of all linear continuous operators in a Hilbert space.

9.68 Define the operator S on $L^2(\mathbb{R})$ by $(Sf)(t) = \exp(-t^2) f(t - 1)$. Describe S^*. Find $\|S^n\|$. Show that S is injective, $\text{Ran } S$ is dense and $\sigma(S)$ is a singleton. Show that if T is defined on $\text{Ran } S$ by $TSf = f$, then $\sigma(T) = \varnothing$.

9.69 Prove that each operator on $L^2(0, 1)$ with finite dimensional range is the integral operator with kernel in $L^2((0, 1) \times (0, 1))$, but not every compact operator on $L^2(0, 1)$ is of this form.

9.70 Check the compactness of the set $\{f \in L^2(0, 1) : f(x) = \Sigma_{n=1}^\infty a_n \sin(2\pi n x), \Sigma_{n=1}^\infty n |a_n| \leqslant 1\}$ in $L^2(0, 1)$.

9.71 Prove the Rellich compactness criterion in $L^2(\mathbb{R})$: If F and G are real functions on $\mathbb{R}^n$ such that $F(x)$, $G(x) \to \infty$ when $|x| \to \infty$, then the set $\{f : \int |f|^2 \leqslant 1, \int F|f|^2 \leqslant 1, \int G|\hat{f}|^2 \leqslant 1\}$ is compact in $L^2(\mathbb{R}^n)$.

9.72 T is a compact selfadjoint operator in a Hilbert space H. Show that if $Tx \neq 0$ for some $x \in H$, then $T^n x \neq 0$ for all n and the sequence of quotients $\|T^{n+1}x\| / \|T^n x\|$ converges to the absolute value of an eigenvalue of T.

9.73 Show the compactness of the Volterra integral operator on $L^2(0,1)$ defined by $Tf(x) = \int_0^x f(t)\, dt$. Moreover $\|T\| \leqslant 2^{-1/2}$, $T + T^*$ is a projection from L^2 onto $\mathbb{C}$, $\sigma(T) = \{0\}$.

9.74 Define the operators S and M in ℓ^2 by $Sx(n) = 0$ if $n = 0$ and $x(n-1)$ if $n \geqslant 1$ and $Mx(n) = x(n)/(n+1)$. Show that $T = MS$ is a compact quasinilpotent operator, that is, $\lim_{n \to \infty} \|T^n\|^{1/n} = 0$. Generalize this result to the case when the sequence $(1/(n+1))$ in the definition of M is replaced by a sequence (c_n) of nonzero reals converging to zero.

9.75 Define the operator T in a Hilbert space H equipped with an orthonormal basis (e_n) by $Te_1 = 0$, $Te_n = e_{n-1}/n$ for $n > 1$. Show that there exists no continuous operator S such that $S^2 = T$.

9.76 Verify that the operator T on $L^2(0, \infty)$ (and more generally on $L^p(0, \infty)$, $1 < p < \infty$) defined by $Tf(t) = t^{-1} \int_0^t f(s)\, ds$ is bounded but not compact.

9.77 Compute the norm of the operator in $L^2(0, 1)$ defined by $Tf(t) = t^{-1} \int_0^t f(s)(1 - s)^{-1}\, ds$.

9.78 Prove that the algebra of all compact operators on a Hilbert space does not have any proper closed two-sided ideal.

9.79 The Halmos theorem says that in infinite dimensional separable complex Hilbert space every contraction is the weak limit of a sequence of unitary operators. Describe the weak sequential closure of the class of operators similar to unitary ones.

9.80 Determine when the convolution with a function $f \in L^1(G, m)$ on a locally compact abelian group G equipped with the Haar measure m is
  (i) selfadjoint
  (ii) unitary
  (iii) compact operator in $L^2(G, m)$

9.81 Give the conditions for
  (i) invertibility
  (ii) compactness

of the convolution operator $Kf(x) = \int_{\mathbb{R}} k(x - y)f(y)\,dy$ on $L^2(\mathbb{R})$, $k \in L^1(\mathbb{R})$.
Show that if the support of k is compact, then K has no eigenvalues.

9.82 Consider the convolution operator on $L^2(\mathbb{R})$ defined by
$Kf(x) = \int_{\mathbb{R}} k(x - y)f(y)\,dy$, $k \in L^1(\mathbb{R})$, and the operators

$$K_a f(x) = \begin{cases} \int_{-a}^{a} k(x - y)f(y)\,dy & \text{for } |x| \leqslant a \\ 0 & \text{for } |x| > a \end{cases}$$

Prove that, except for some trivial cases, K_a does not converge to K in the
operator norm if a tends to infinity. Assume that $\|k\|_1 < 1$ and consider the
equations $f_a - K_a f_a = g$, $g \in L^2(\mathbb{R})$, satisfied for $|x| \leqslant a$ and $f_\infty - K f_\infty = g$.
Prove that f_a are quite good approximations of f_∞:

$$\|(f_a - f_\infty)_{(-a,a)}\|_2 \leqslant (1 - \|k\|_1)^{-1} \|k\|_1 \left(\int_{|x| > a} |f_\infty|^2 \right)^{1/2}$$

9.83 Give an example of an infinite matrix (a_{ij}) such that for $i, j = 1, 2, \ldots$
$\sum_{i=0}^{\infty} |a_{ij}|^2 \leqslant 1$, $\sum_{j=0}^{\infty} |a_{ij}|^2 \leqslant 1$ and the operator A defined by $(Ae_i, e_j) = a_{ij}$ is
not continuous on ℓ^2.

9.84 Let A be the operator on ℓ^2 defined by an infinite matrix (a_{ij}) such that
for some $k > 0$ $a_{ij} = 0$ if $|i - j| > k$ and $\lim_{i,j \to \infty} a_{ij} = 0$. Show that A is
compact.

9.85 Does the matrix

$$\begin{bmatrix} 1 & 1/2 & 1/3 & 1/4 & 1/5 & 1/6 & \cdots \\ 0 & 1 & 0 & 1/2 & 0 & 1/3 & \cdots \\ 0 & 0 & 1 & 0 & 0 & 1/2 & \cdots \\ \hdashline & & & & & & \end{bmatrix}$$

define a bounded operator on ℓ^2?

9.86 Prove the Hilbert inequality $\sum_{i,j=1}^{n} a_i b_j/(i + j) \leqslant \pi (\sum_{i=1}^{n} a_i^2)^{1/2}$
$(\sum_{j=1}^{n} b_j^2)^{1/2}$ for arbitrary $a_i, b_j \in \mathbb{R}$.
  Show that the infinite Hilbert matrix

$$\begin{bmatrix} 1 & 1/2 & 1/3 & 1/4 & \cdots \\ 1/2 & 1/3 & 1/4 & \cdots \\ 1/3 & 1/4 & \cdots \\ 1/4 & \cdots \\ \cdots \end{bmatrix}$$

defines on ℓ^2 a linear operator of the norm equal to π.

9.87 Show that the infinite matrix (a_{ij}) where $a_{ij} = \exp(-i - je^{-2i})$ defines on ℓ^2 the bounded operator.

9.88 Define the infinite Kolmogorov matrix (a_{ij}) by $a_{ii} = -i$, $a_{i,i+1} = 1$ for $i \geqslant 1$, $a_{i,i-1} = i - 1$ for $i > 1$ and $a_{ij} = 0$ otherwise. Show that this matrix defines the bounded operator A in the space of complex sequences $x = (x_n)$ equipped with the norm $\|x\| = \Sigma_{n=1}^{\infty} |x_n|/(2n)!$. The equation $Ax = \lambda x$ has a solution such that $x_n = x_n(\lambda)$ which is a polynomial in λ of degree $n - 1$ and $x_n(0) = 1$, $|x_n(\lambda)| \leqslant n!|\lambda|^{n-1}$ for $|\lambda| \geqslant 1$. Describe the discrete spectrum of A.

9.89 Consider the Hardy space H^2 of analytic functions $f(z) = \Sigma_{n=0}^{\infty} c_n z^n$ in the unit disk in $\mathbb{C}$ such that $\|f\|^2 = \Sigma_{n=0}^{\infty} |c_n|^2 < \infty$. Show that the (bounded) operator V on H^2 defined by $(Vf)(z) = zf(z)$ is the Cayley transform of the symmetric operator $(Tf)(z) = i(1 + z)(1 - z)^{-1}f(z)$. Find $\mathrm{Ran}(T \pm iI)$. Study in a similar manner $(Vf)(z) = zf(z^2)$.

9.90 Let $H(D)$ denote the vector space of the holomorphic functions in a domain $G \subseteq \mathbb{C}$, continuous in $\bar{G}$. Show that the multiplication operator associated to a function g holomorphic in a domain containing $\bar{G}$ is a Fredholm operator. Compute its index.

9.91 Give an example of a nonlinear operator on ℓ^2 which is continuous in the norm but unbounded on the balls of sufficiently large radii.

9.92 Construct an example of nonlinear operator on the unit ball in ℓ^2 without fixed point.

9.93 Prove that the recurrent sequence $X_0 = 0$, $X_{n+1} = X_n + \frac{1}{2}(A - X_n^2)$, where $0 \leqslant A \leqslant I$ is a positive operator, strongly converges to a (unique) positive operator X such that $X^2 = A$.

9.94 Prove (without the spectral theorem) that every positive bounded operator has (positive, bounded) roots of any order.

9.95 Characterize the dual space for the space of all nuclear operators on a Hilbert space as the space of all bounded operators. The mapping $A \mapsto \mathrm{Tr}(A \cdot)$ establishes this duality. Then show that the space of nuclear operators is the dual space for the space of all compact operators.

9.96 Solve the Dirichlet problem for the Laplacian in a simply connected region in $\mathbb{R}^n$ with continuous boundary condition using the Hahn-Banach theorem.

Then observe that having this solution it is easy to construct the Green function and the conformal mapping of this region onto the unit disk (that is, to prove the Riemann mapping theorem).

9.97 Construct in an infinite dimensional separable Hilbert space a densely defined operator T such that $D(T^*) = \{0\}$.

9.98 Give an example of densely defined linear operator on a Hilbert space which is not closable.

9.99 Show that for every densely defined closed operator A, $I + A^*A$ is selfadjoint.

9.100 The operators A and B are positive on a Hilbert space. Show that $I + AB$ is invertible.

9.101 Consider the subset $\ell^2 \supseteq D(A) = \{x = (x_n) \in \ell^2$: there exists N such that $\Sigma_{n=0}^{N} x_n = 0$ and $x_n = 0$ for $n > N\}$ and define the linear operator A on $D(A)$ by $(Ax)_n = i(\Sigma_{m=0}^{n-1} x_m + \Sigma_{m=0}^{n} x_m)$. Prove that $D(A)$ is dense in ℓ^2, A is symmetric, $\text{Ran}(A + iI)$ is dense in ℓ^2, but A has no selfadjoint extensions.

9.102 Define for the operator T on a Hilbert space the operators $K_n = I + \Sigma_{k=1}^{n} (1 - k/(n + 1))(T^k + T^{*k})$. Show that if T is a contraction, then $K_n \geqslant 0$ (an operator analog of the Fejér kernel).

9.103 T is a bounded linear operator on a Hilbert space with the spectral radius equal to 1. Show that for any $c > 1$ there exists an invertible operator A such that $\|ATA^{-1}\| \leqslant c$.

9.104 The operators P and Q are the orthogonal projections on a Hilbert space such that $P - Q$ is a nuclear operator. Prove that $\text{Tr}(P - Q)$ is an integer.

9.105 Prove the Calkin theorem: Each symmetric operator S in a Hilbert space satisfying the inequality $\|Sf\| \geqslant \|f\|$ on its dense domain has a selfadjoint extension with the same property.

Generalize to the case of the symmetric operator S such that
$$\left\|\left(S - \frac{a+b}{2} I\right) f\right\| \geqslant \frac{b - a}{2} \|f\|$$ for some $a < b$ (the interval (a, b) is then a lacuna in the spectrum of A).

9.106 Represent each unitary operator U on a Hilbert space as $U = \exp(iH)$ with a bounded selfadjoint H. In particular find such representation for the shift in $\ell^2(\mathbb{Z})$ $(Ux)(n) = x(n + 1)$ with $\|H\| \leqslant \pi$. Study the uniqueness of this representation.

9.107 Let $A = U|A|$ be the polar decomposition of the operator A. Show that U is the strong limit of a sequence of some functions of A (that is U is in the von Neumann algebra generated by A).

9.108 Let A be a bounded operator on a Hilbert space, C a closed Jordan curve disjoint from the spectrum of A. Show that if $I = (2\pi i)^{-1} \oint_C (zI - A)^{-1} dz$, then the spectrum of A is contained in the region bounded by C.

9.109 A is a densely defined closed operator on a Hilbert space, Ran $A \subseteq D(A)$, $A^2|_{D(A)} = I$. Does it imply that $\sigma(A) \subseteq \{-1, 1\}$?

9.110 Let P and Q be projections (not necessarily orthogonal) on a Hilbert space such that $\|P - Q\| < 1$. Define $A = (I - P)(I - Q) + PQ$, $B = (I - Q)(I - P) + QP$, $C = I - (P - Q)^2$. Show that $AB = BA = C$. Then if $(1 - x)^{-1/2} = \Sigma_{n=0}^{\infty} a_n x^n$ is the Taylor series, for $D = \Sigma_{n=0}^{\infty} a_n (P - Q)^n$ the equalities $D^2C = CD^2 = DCD = I$ hold. Show that $P(P - Q)^2 = (P - Q)^2 P$, $Q(P - Q)^2 = (P - Q)^2 Q$, $DP = PD$, $QD = DQ$. Putting $W = DA$ one has $W^{-1} = BD$, $WQ = PW$. Finally show that if P, Q are selfadjoint, then the intertwining operator W is unitary.

9.111 Prove the Weyl-von Neumann theorem: If H is a selfadjoint operator in a separable Hilbert space, then there exists a compact selfadjoint operator K such that the eigenvectors of $H + K$ form an orthonormal basis of the whole space.

9.112 Prove that for every unitary operator U in a separable Hilbert space there exists a compact operator K such that the eigenvectors of $U + K$ form an orthonormal basis of the space.

9.113 Let $H_1 = L^2(\mathbb{R})$ and H_2 be the closure of $\mathscr{S}(\mathbb{R})$ in H_1 in the norm $\|f\|_2 = (\|xf\|^2 + \|f'\|^2)^{1/2}$. The operators $A_{\pm} : H_2 \to H_1$ (the creation and annihilation operators in quantum field theory) are defined by $A_{\pm} f = (d/dx \pm x)f$. Show that $A_{\pm}$ are Fredholm operators and compute their indexes.

9.114 $\{S(t) : t \geq 0\}$ is a strongly continuous semigroup of bounded selfadjoint operators in a Hilbert space. Show that $S(ct + (1 - c)s) \leq cS(t) + (1 - c)S(s)$ for $0 \leq c \leq 1$, $t, s > 0$.

9.115 It is well-known that for the matrices A and B $\exp(A + B) = \lim_{n \to \infty} (\exp(A/n) \exp(B/n))^n$. Determine the behavior of the operators $(\exp(At/n) \exp(Bt/n))^n$, $0 < t < \infty$, for $n \to \infty$, where A, B are the operators on $L^2(\mathbb{R})$ defined as follows: $Af = f'$, $B = VAV^*$, where V is the operator of multiplication by the function

   (i) $2\chi_{E+\mathbb{Z}}(x) - 1$ for any set $E \subseteq (0, 1)$ such that $0 < m(E \cap (a, b)) < b - a$ for all $0 \leq a < b \leq 1$.
   (ii) $\exp(ig(x))$ for a real C^1 function g.

Remark: for (i) see problems 6.11–6.13.
Under some additional hypotheses on the operators A and B which are the infinitesimal generators of the strongly continuous semigroups $\exp(At)$, $\exp(Bt)$, the sequence $(\exp(At/n) \exp(Bt/n))^n$ strongly converges to $\exp((A + B)t)$, see the Lie-Trotter formula, for example $[R - S]$ VIII.8, X.51.

9.116 Let A and B be selfadjoint operators in a Hilbert space. Prove that $\|e^{iA} - e^{iB}\| \leqslant \|A - B\|$.

9.117 Given any infinite dimensional separable normed space construct a countable dense set which is linearly independent.

9.118 (x_n) is a sequence of nonzero vectors in a normed space X. Show that there exists a closed subspace of codimension 1 which does not contain any x_n.

9.119 An infinite matrix (a_{ij}) is said to be reversible if for every convergent sequence (y_n) there exists a unique sequence (x_n) such that $\Sigma_{n=0}^{\infty} a_{in}x_n = y_i$, $i = 0, 1, 2\ldots$. Find two reversible matrices such that their product is not reversible.

9.120 Compute the distance in $C[-1,1]$ from the function x^n to the subspace of the polynomials of degree less than n.

9.121 Does there exist a function $f \in C^{\infty}[0,1]$ such that $\text{lin}\{f^{(n)} : n \in \mathbb{N}\}$ is dense in $C[0,1]$?

9.122 The formula $(Sf)(x) = (f(x/2) - f(1 - x/2))/2$ defines an operator on $C[0,1]$. Show that there exists a unique linear functional h on $C[0,1]$ such that $h(f) \in f([0,1])$ and $h(Sf) = h(f)$ for all f.

9.123 Verify that $h(f) = \Sigma_{n=0}^{\infty} (-2)^{-n} f(q_n)$, where the sequence (q_n) contains all the rationals in $[0,1]$, is a continuous linear functional on $C[0,1]$. Show that the norm of h is not attained on the unit ball.

9.124 Prove that $C[0,1]$ over reals is not a dual space (for any normed space). The same for c, c_0, $L^1(0,1)$.

9.125 Show that the set $\{f : \Sigma_{n \in \mathbb{Z}} |\hat{f}(n)|(1 + |n|) \leqslant 1\}$ is relatively compact in $C[0, 2\pi]$.

9.126 A linear operator $T : C(\mathbb{R}) \to C(\mathbb{R})$ is said to be local if the equality $f_1 = f_2$ on $[a,b]$, $a < b$, implies $Tf_1 = Tf_2$ on $[a,b]$. Show that such an operator is of the form $Tf(x) = g(x)f(x)$ for some continuous function g.

9.127 X is a subspace of $C^1[0,1]$ closed in $C[0,1]$. Show that X is finite dimensional (hence closed) subspace in $C^1[0,1]$.

9.128 Construct a Schauder basis in $C[0,1]$.

9.129 Define for any continuous function $f : \mathbb{R} \to \mathbb{R}$ $\|f\| = \sup_x \Sigma_{n \in \mathbb{Z}} |f(x + n)|$. Prove that the set $\{f : \|f\| < \infty\}$ equipped with the norm $\|\cdot\|$ is a Banach space.

9.130 Prove that the space of the (normalized) functions of bounded variation $BV_0[0, 1] = \{f \in BV[0, 1] : f(0) = 0\}$ is a Banach algebra with pointwise multiplication.

Corollary: if f and g are absolutely continuous on $[0, 1]$, $f(0) = g(0) = 0$, then $\int_0^1 |(fg)'| \leqslant \int_0^1 |f'| \int_0^1 |g'|$.

9.131 Construct an example of a measure space (M, μ) such that $(L^1)^* \neq L^\infty$.

9.132 The sequence $(x_n) \subseteq L^p$, $1 < p < \infty$, is such that for some $y \in L^p$ the limit $\lim_{n \to \infty} \|x_n - ty\|$ exists for all $t \in [0, 1]$. For which p does it imply the existence of $\lim_{n \to \infty} \|x_n - ty\|$ for all $t \geqslant 0$?

9.133 Consider the equality $\|x - y\| \, \|x + y\| = |\, \|x\|^2 - \|y\|^2 |$ in ℓ^p over reals, $1 < p < \infty$. Does it imply that $y = cx$ for some $c \in \mathbb{R}$?

9.134 Prove the Clarkson inequalities
(i) For $p \geqslant 2$, $f, g \in L^p$

$$\|(f + g)/2\|^p + \|(f - g)/2\|^p \leqslant (\|f\|^p + \|g\|^p)/2.$$

(ii) For $1 < p < 2$, $f, g \in L^p$, $1/p + 1/p' = 1$

$$\|(f + g)/2\|^{p'} + \|(f - g)/2\|^{p'} \leqslant ((\|f\|^p + \|g\|^p)/2)^{1/(p-1)}.$$

9.135 Prove the Radon-Riesz theorem: If $f_n \to f$ weakly in L^p, $1 < p < \infty$, and $\|f_n\| \to \|f\|$, then $\|f_n - f\| \to 0$.

9.136 $X \subseteq \ell^1$ contains all the vectors with even coordinates equal to zero, $Y \subseteq \ell^1$ contains (x_n) such that $x_{2n} = 2^{-n} x_{2n-1}$. X and Y are both closed subspaces. Is $X + Y$ also closed?

9.137 $Y \subseteq L^1(\mathbb{R}^n)$ is a closed subspace. Show that Y is translation invariant if and only if $f * g \in Y$ for all $f \in Y$, $g \in L^1(\mathbb{R}^n)$.

9.138 Show that if $f \in L^1(\mathbb{R}^+)$, g is measurable and $|g(t)/t| \leqslant M < \infty$ for $t \geqslant 1$, then $\lim_{t \to \infty} t^{-1} \int_1^t f(s)g(s) \, ds = 0$.

9.139 Define for $f \in L^1(0, 1)$ the operator $(Tf)(x) = x^{-1} \int_0^1 f(t) \, dt$. Give the examples of bounded continuous functions on $(0, 1]$ such that
(i) f does not tend to any limit at 0, Tf does.
(ii) Tf does not have any limit at 0.
(iii) $T^n f$ does not have any limit at 0 for all $n \in \mathbb{N}$.

9.140 Construct examples of infinite dimensional closed subspaces of $L^1(\mathbb{R})$
(i) containing continuous functions only
(ii) without any nonzero continuous function

9.141 Show that the sequence of functionals (δ_n) on the space ℓ^∞ such that $\delta_n((x_k)) = x_n$ does not have any weak* convergent subsequences but it has a weak* convergent subnet.

9.142 X is a closed subspace of $L^1(0,1)$ such that $X \subseteq \bigcup_{p>1} L^p(0,1)$. Show that $X \subseteq L^r(0,1)$ for some $r > 1$.

9.143 Prove the Oxtoby theorem: $L^2(0,1) \subseteq L^1(0,1)$ is the first category subset. More generally: $L^p(0,1) \subseteq L^q(0,1)$ for $1 \leqslant q < p$ is also of the first category.

9.144 The Banach mean is any linear continuous functional L on ℓ^∞ such that $L(x_0, x_1, \ldots) \geqslant 0$ if $x_n \geqslant 0$, $L(x_0, x_1, x_2, \ldots) = L(x_1, x_2, x_3, \ldots)$ and $L(1, 1, 1, \ldots) = 1$. Prove that
  (i) There exists a Banach mean.
  (ii) $\lim_{n \to \infty} (\inf_j n^{-1} \sum_{i=0}^{n-1} x_{i+j}) \leqslant L(x_0, x_1, \ldots) \leqslant \lim_{n \to \infty} (\sup_j n^{-1} \sum_{i=0}^{n-1} x_{i+j})$.
  (iii) The necessary and sufficient condition for the coincidence of the values of all Banach means on a sequence (x_n) is that $\lim_{n \to \infty} n^{-1} \sum_{i=0}^{n-1} x_{i+j} = s$ uniformly with respect to j, $s = L((x_n))$.

9.145 h is a linear functional on the algebra ℓ^∞ (a priori not necessarily continuous) such that for every $x \in \ell^\infty$ a subsequence of x converges to $h(x)$. Is h multiplicative, that is, $h(xy) = h(x)h(y)$ for all x and y?

9.146 Prove that each bilinear continuous form on $L^1(\mathbb{R})$ is determined by a function $F \in L^\infty(\mathbb{R}^2)$: $F(f,g) = \int F(x,y)f(x)g(y)\,dxdy$. Give examples of bilinear continuous forms on $L^p(0,1)$, $1 < p < \infty$, which are not of this form for any $F \in L^q(\mathbb{R}^2)$, $1/p + 1/q = 1$.

9.147 Show that the closure of the set of translates $\{f(x-t): t \in \mathbb{R}\}$ of the function $f(x) = \exp(ix^2)$ in the topology of pointwise convergence on the space of all bounded complex functions defined on $\mathbb{R}$ contains a non-measurable function.

9.148 Consider the set $C(X)$ of all continuous functions on a completely regular topological space X equipped with the topology of uniform convergence on compact sets in X. Prove that for each linear continuous functional $h \not\equiv 0$ on $C(X)$ there exists the smallest compact K such that $f \equiv 0$ on K implies $h(f) = 0$.

9.149 Determine the domain $D(A)$ of the operator $A = d^2/dt^2$ in such a way that $A: D(A) \to X = \{f \in C[0,1]: f(0) = f(1) = 0\}$ is densely defined and closed. Then show that for each function $f \in D(A)$ there exists a functional $F = F_f \in X^*$ such that $\|F(f)\| = \|f\|^2$, $F(Af) \leqslant 0$.

9.150   Consider the operators $T_n f(t) = n(f(t + 1/n) - f(t))$, $n = 1, 2, \ldots$, defined in the space $C(\mathbb{T})$ of continuous 2π-periodic functions. Prove that the set of f such that $\lim_{n \to \infty} T_n f(t) = f'(t)$ exists for all $t \in \mathbb{T}$ is dense in $C(\mathbb{T})$, $\|T_n\| = 2n$ and $\lim_{n \to \infty} T_n f$ cannot exist on any subset of the second Baire category in $C(\mathbb{T})$.

Corollary: the set of nowhere differentiable functions is of the second category in $C(\mathbb{T})$ (in fact its complement is of the first category).

9.151   Consider the family of the Picard transforms in $C(\mathbb{R})$ or $L^p(\mathbb{R})$ defined by $T(t)f(x) = \frac{t}{2} \int_{\mathbb{R}} \exp(-t|x - u|) f(u) \, du$, $t > 0$. Show that
   (i) $T(t)$ is a contraction.
   (ii) $\lim_{t \to \infty} T(t)f(x) = f(x)$ uniformly on compacts in $\mathbb{R}$ for all $f \in C(\mathbb{R})$.
   (iii) The functions $F_t(x) = T(t)f(x)$ satisfy the equation $F_t''(x) - t^2 F_t = -t^2 f(x)$.
   (iv) The following identity holds:

$$(s^2 - t^2)T(s)(T(t)f) = s^2 T(t)f - t^2 T(s)f \qquad s, t > 0$$

   (v) if f is measurable and $T(t)f = f$ for all $t > 0$, then $f(x) = ax + b$ for some a and b.

9.152   It is well-known that if A is the infinitesimal generator of a strongly continuous semigroup of contractions on a Banach space X, then for $x \in D(A^2)$ the Kallman-Rota inequality holds: $\|Ax\|^2 \leqslant 4\|x\| \|A^2 x\|$. Deduce from this
   (i) the classical Landau inequality

$$\|f'\|^2 \leqslant 4\|f\| \|f''\| \quad \text{for } f \in C^2[0, \infty], \; X = C[0, \infty]$$

   (ii) the S. Kurepa inequality

$$\|f''\|^2 \leqslant \frac{4}{3} \|f\| \|f^{(4)}\| \quad \text{for } f \in C^4[0, \infty]$$

Prove the Kallman-Rota inequality.

9.154   Consider a set of couples $B = \{(x_\alpha, y_\alpha) : \alpha \in A\} \subseteq X \times X^*$, where X is a Banach space and $y_\beta(x_\alpha) = \delta_{\alpha\beta}$. B is called a generalized basis if $\text{lin}(x_\alpha)$ is dense in X and a dual generalized basis if $\text{lin}(y_\beta)$ is dense in X^*. Do these definitions coincide if one considers as X a Hilbert space (and X^* identified canonically with X)?

9.155   A sequence (X_n) of finite dimensional subspaces of an infinite dimensional Banach space X is called the Schauder decomposition of X if every element $x \in X$ has a unique representation $x = \sum_{n=1}^\infty x_n$ with $x_n \in X_n$. It is well-known that $v((X_n)) = \sup_{k \in \mathbb{N}} \|P_k\| < \infty$, where $P_k(x) = \sum_{n=1}^k x_n$. Show that every infinite dimensional Banach space contains a subspace such that for each $\varepsilon > 0$ there exists its Schauder decomposition (X_n) with $v((X_n)) < 1 + \varepsilon$.

9.156 It is well-known that if X is a Banach space, then a subset of X^* is weak* compact if and only if it is weak* closed and bounded in the norm.
  (i) Show that this is not true in general in normed spaces.
  (ii) Prove that for a normed space X a weak* compact subset of X^* is bounded in the norm if and only if the weak* closure of its convex hull is weak* compact.

9.157 Consider a nontrivial linear continuous functional h on an infinite dimensional normed space. Is it possible that the image $h(C)$ of every bounded closed subset C is compact in the field of scalars?

9.158 Consider extensions of linear continuous functionals from codimension one subspace of a given Banach space given by the Hahn-Banach theorem (that is, those with nonincreasing norm). When are they unique?

9.159 Prove the Banach and Steinhaus uniform boundedness principle without use of the notion of Baire category, that is more precisely, for a family of operators A_n from a Banach space X into a normed space Y such that $\sup_n \|A_n\| = \infty$ construct a vector x satisfying the condition $\|A_n x\| \geqslant n$ for all n.

9.160 Prove that each separable Banach space is isometrically isomorphic to a quotient space ℓ^1/Y for some closed subspace $Y \subseteq \ell^1$.

9.161 Show that each bounded linear operator from a reflexive Banach space into ℓ^1 is compact.

9.162 T is a bounded linear operator on a Banach space X. Must the closure of the image $T(B)$ of the unit ball in X be contained in the range of T?

9.163 Let X, Y, and Z be Banach spaces, $T: X \to Z$, $S: Y \to Z$ linear continuous operators. Show that the (equivalent) conditions
  (i) $\overline{T(B_1)} \supseteq S(B_\varepsilon)$ for some $\varepsilon > 0$ (B_ε is the ball of the radius ε centered at the origin).
  (ii) $\|S^*f\| \leqslant k\|T^*f\|$ for all $f \in Z^*$ and some $k > 0$.
do not imply $T(X) \supseteq S(Y)$.

9.164 Prove the Banach-Schauder theorem: If $T: X \to Y$ is a linear continuous operator between the Banach spaces X and Y, then $T(U)$ is open for every open set $U \subseteq X$ or $T(X)$ is of first category in $\overline{T(X)}$.

9.165 Show that the kernel of a compact linear operator in a Banach space need not be the range of a continuous projection (that is, need not be complemented).

9.166 T is a compact operator in a Banach space. Is the image of the unit ball $T(B)$ compact?

9.167 Prove that a bounded linear operator in a reflexive Banach space is compact if and only if the image of every weakly convergent sequence is a norm convergent sequence.

9.168 A and B are bounded linear operators in a Banach space. Show that if B is compact and Ran $A \subseteq$ Ran B, then A is also compact.

9.169 K is a compact (Hausdorff) space, A is a subalgebra of the (noncommutative) algebra $C(K, \mathbb{H})$ of the quaternion valued continuous functions on K that separates the points of K and contains the constants. Show that A is dense in $C(K, \mathbb{H})$.

9.170 A is a commutative Banach algebra with 1, M, and N its maximal ideals. Show that $S = \lin\{mn : m \in M, n \in N\}$ is of finite codimension if $M \neq N$ (this is not necessarily the case if $M = N$).

9.171 Solve the differential equation $dX/dt = -XAX$, where $X(t)$, A are bounded linear operators on a Banach space, A is fixed, and $X(0) = I$.

9.172 Let (C_n) be a decreasing sequence of bounded closed connected subsets of a Banach space. Is the intersection $\bigcap_{n=1}^{\infty} C_n$ connected?

9.173 Consider a nonlinear operator T defined on a compact convex subset C of a Banach space which is the contraction: $\|T(x) - T(y)\| \leqslant \|x - y\|$ for $x, y \in C$. Prove in an elementary way (without recourse to the Schauder fixed point theorem) that T has a fixed point.

9.174 T is a nonlinear operator in a normed space X such that $\|T(x) - T(y)\| \leqslant \|x - y\|$, $\|T(x) - T^2(x)\| < \|x - T(x)\|$ for $x \neq T(x)$ and the sequence $(T^n(x))$ has a cluster point. Prove that for all x the iterates $T^n(x)$ converge to a fixed point of T.

9.175 What can be said about a function f belonging to
  (i) $\mathscr{S}(\mathbb{R})$
  (ii) $\mathscr{D}(\mathbb{R})$
if all its moments $\int x^n f(x)\, dx$ vanish?

9.176 Show that if $\Sigma_{n=0}^{\infty} f^{(n)}(x) = 0$ for a smooth function f and this series converges uniformly on compacts, then $f \equiv 0$.

9.177 A topology in the space $C_c(\mathbb{R})$ of the continuous functions with compact support in $\mathbb{R}$ is defined by the neighborhoods of 0 of the form $\{f \in C_c : |f(x)| \leqslant F(x)\}$, where $F : \mathbb{R} \to (0, \infty)$ is any positive continuous function.

Show that the sequence (f_n) converges to 0 in this topology if and only if the supports of f_n are contained in a fixed compact and f_n converge to 0 uniformly. Prove that this topology is not metrizable. Show that 0 is in the closure of the set $\{f(x)/m + f(x/m)/k : m, k \in \mathbb{N}\}$, for every $f \in C_c$, $f(0) \neq 0$, but 0 is not a limit of any sequence of the elements from this set.

9.178   Show that if the derivatives $D_1 u, \ldots, D_n u$ of a distribution u on $\mathbb{R}^n$ are in L^2, then u is locally in L^2. For $n = 1$, u is even continuous. What can be said about the case $n = 2$?

9.179   For a function $f \in \mathscr{S}$ define the translate f_y by the formula $f_y(x) = f(x - y)$. Then for tempered distribution $T \in \mathscr{S}'$ denote by T^f the function (!) such that $T^f(y) = T(f_y)$. Show that $T^f \in \mathscr{S}$. Assume that $f_n \in \mathscr{S}$, $f_n \to \delta$ in $\mathscr{S}'$. Does it imply the convergence $T^{f_n} \to T$ in $\mathscr{S}'$ for every $T \in \mathscr{S}'$?

9.180   Give an example of a tempered distribution T defined by a smooth function f of polynomial growth such that the set of translates $\{T_x : x \in \mathbb{R}\}$ is bounded as a set of distributions but f is unbounded.

9.181   Prove that the function $\exp(ie^x)$ is a tempered distribution but its derivative in $\mathscr{S}'$ is not equal to $ie^x \exp(ie^x)$.

9.182   Construct an example of distribution in $\mathscr{D}'(0, \infty)$ with no extension to $\mathscr{D}'(\mathbb{R})$.

9.183 P is a polynomial of degree n, $a_1, \ldots, a_k$ are its real roots of multiplicities $m_1, \ldots, m_k$ respectively. Solve in $\mathscr{D}'(\mathbb{R})$ the equations $PT = 0$ and $PT = S$ where S is a distribution with its support disjoint from $\{a_1, \ldots, a_k\}$.

9.184   Compute the limits in $\mathscr{D}'(\mathbb{R})$

$$\lim_{a \to 0+} a/(x^2 + a^2), \quad \lim_{a \to 0-} a/(x^2 + a^2)$$

$$\lim_{a \to 0+, h \to 0} a(2h)^{-1}/(((x - h)^2 + a^2)^{-1} - ((x + h)^2 + a^2)^{-1})$$

9.185   Compute in $\mathscr{D}'(\mathbb{R})$

$$\lim_{a \to 0} a^{-1}(\text{p.v. } (x + a)^{-1} - \text{p.v. } (x - a)^{-1})$$

9.186   Define the differential operators in $\mathscr{S}'$ $D_\varepsilon u = -\varepsilon(d^2 u/dx^2) + du/dx + u$, $\varepsilon \geq 0$. Solve the equation $D_\varepsilon u = \delta$. Study convergence of the (unique !) solutions u_ε as $\varepsilon \to 0$.

9.187   Consider the regular distributions $f_{a,v}(x) = a^v \exp(ax^2)$, $g_{a,v,p}(x) = a^v x^p \exp(ax^2)$, where $a \in \mathbb{C}$, $v \in \mathbb{R}$, $v > -1$, $p \in \mathbb{N}$. Study the convergence of $f_{a,v}$, $g_{a,v,p}$ in $\mathscr{D}'$ when $|a|$ tends to infinity.

9.188 Show that for a given function f locally integrable in $\mathbb{R}\backslash\{0\}$ having power singularity at the origin, there exists a distribution in $\mathscr{D}'$ (also denoted by f) such that $(f, \varphi) = \int f(x)\varphi(x)\,dx$ for all test functions φ vanishing in a neighborhood of 0.

Moreover show that if $|f(x)| > c_m/|x|^m$ for every m and some constants c_m, $0 < |x| < \varepsilon$, then this problem of regularization of a function to a distribution does not have any solution.

9.189 Show that the equation $-2x^3 y' = y$ has no nontrivial solution in $\mathscr{D}'$.

9.190 Show that the differential equation $xy' = f$ has a distributional solution for every $f \in \mathscr{D}'$.

9.191 Solve the functional equation in $\mathscr{D}'$ $(T, f * g) = (T, f)(T, g)$ for all $f, g \in \mathscr{D}$.

What can be said about $S \in \mathscr{D}'$ satisfying $(S, fg) = (S, f)(S, g)$ for all $f, g \in \mathscr{D}$?

9.192 Characterize the complex sequences (a_n) such that $(a_n \delta_n)$ converges to zero in $\mathscr{D}'$, in $\mathscr{E}'$, in $\mathscr{S}'$.

9.193 Study convergence in $\mathscr{D}'$, $\mathscr{S}'$, $\mathscr{E}'$ of the distributional series $\Sigma_{n=0}^{\infty} a^n \delta_n$, $\Sigma_{n \in \mathbb{Z}} a^n \delta_n$ for $a > 0$. Compute the Fourier transforms of these series.

9.194 Characterize the real sequences (s_n) such that the series $T = \Sigma_{n=0}^{\infty} s_n \delta_n$ converges in $\mathscr{D}'$, $\mathscr{S}'$. Prove that if $s_0 \neq 0$, then there exists the unique distribution E supported on $\mathbb{R}^+$ such that $T * E = \delta$. Is E the tempered distribution?

9.195 Determine the conditions for convergence of the series $\Sigma_{n \in \mathbb{Z}} c_n e^{2\pi i n x}$ in $\mathscr{D}'$.

9.196 Show that if $f \in L^{\infty}(\mathbb{R}^n)$ and the support of $\hat{f}$ is finite and equal to $\{s_1, \ldots, s_k\}$, then f is a trigonometric polynomial of the form $f(x) = a_1 \exp(is_1 x) + \cdots + a_k \exp(is_k x)$ a.e.

9.197 A distribution $f \in \mathscr{D}'$ converges to 0 at infinity if $\lim_{|h| \to \infty} (f, \varphi(x + h)) = 0$ for all test functions φ. Verify that $f(x) = d((\sin^2 x)/x)/dx$ has this property. Show that a polynomial convergent to 0 in this sense vanishes identically (similarly as analytic functions of moderate growth).

9.198 A distribution $f \in \mathscr{D}'$ has polynomial growth at ∞ if $(f, \varphi(x + h))$ is of polynomial growth as the function of h for all test functions φ. Show that such f is a tempered distribution.

9.199 Consider the piecewise continuous functions $u = u(x, t)$ which solve in distribution sense the equation $\partial u/\partial t + \partial F(u)/\partial x = 0$, where $F \in C^1(\mathbb{R})$. Show that on the curve $x = x(t)$ of discontinuities of u the Rankine-Hugoniot condition is satisfied

$$\frac{dx}{dt} = \frac{F(u^+) - F(u^-)}{u^+ - u^-}$$

where u^+, u^- are the right (resp. left) limits of u at the point of discontinuity.

9.200 Assume that $\lim_{n \to \infty} \varphi_n = f$, $\lim_{n \to \infty} \psi_n = g$ in $\mathcal{D}'$ for two sequences (φ_n), (ψ_n) in $\mathcal{D}$. What can be said about the convergence of $(\varphi_n \psi_n)$? Consider the special case $\varphi_n = \psi_n$, $\lim_{n \to \infty} \varphi_n = \delta$.

9.201 For which sequences (a_n) does the series $\Sigma_{n=0}^{\infty} a_n \delta^{(n)}$ converge in $\mathcal{D}'$?

Answers to Chapter 1
Real and Complex Numbers

1.1 **Hint:** observe that $(\sqrt{2^{\sqrt{2}}})^{\sqrt{2}} = 2$.

Remark: it may be shown using the Gelfond theorem that $\sqrt{2^{\sqrt{2}}}$ is transcendent.

1.2 **Answer:** if p_n is the nth prime number, then there is a constant k such that $p_{n+1} - p_n < kp_n^{5/8}$. Let $\delta = 3c/8 - 1 > 0$. If $N > k^{1/\delta}$ and $p_n < N^c < p_{n+1}$, then $p_{n+1} < (N+1)^c - 1$ so there exists a sequence of prime numbers (q_n) such that $q_n^c < q_{n+1} < (q_n + 1)^c - 1$. Now let $u_n = q_n^{c^{-n}}$, $v_n = (q_n + 1)^{c^{-n}}$. The sequence (u_n) increases and (v_n) decreases. Take as θ any number satisfying the inequalities $\lim_{n \to \infty} u_n \leqslant \theta \leqslant \lim_{n \to \infty} v_n$.
AMM 76, 1969, 23–28, U. Dudley.

1.3 **Hint:** See answer to 1.2. Construct the sequences (u_n) and (v_n) using the inequality $p_{n+1} - p_n < p_n$ and the iterated logarithm.
AMM 76, 1969, 23–28, U. Dudley.

1.4 [H-S] 5.40.

1.5 **Hint:** consider separately the cases $\frac{1}{3} \leqslant x \leqslant 1, 0 < x < \frac{1}{3}$. In the first case define $n_0 = 1$, $x_0 = x$, $x_k = n_k x_{k-1}/n_{k-1} - 1$, where

$$n_k = \begin{cases} 4n_{k-1} & \text{if } 1/3 \leqslant x_{k-1} < 1/2 \\ 3n_{k-1} & \text{if } 1/2 \leqslant x_{k-1} < 2/3 \\ 2n_{k-1} & \text{if } 2/3 \leqslant x_{k-1} \leqslant 1 \end{cases}$$

[CHM] 7/1957, N.14.

1.6 Hint: construct the sequences (k_n) and (m_n) by induction so that $k_n a_n \leqslant x - \Sigma_{j=1}^{n-1} k_j a_j < (k_n + 1)a_n$, $m_n a_n \leqslant x/\Pi_{j=1}^{n-1} m_j a_j < (m_n + 1)a_n$.
AMM 87, 1980, 410–411, 6240, M. Eşanu; Le Baron; O. Ferguson.

1.7 See 2.97, 3.31, [G] 39.

1.8 Answer: inf $C = 0$, sup $C = 1$.
AMM 80, 1973, 326–327, 5829, D. P. Giesy; N. Felsinger.

1.9 [H-S] 22.34.

1.10 Hint: show that $n_{k+1} \leqslant 2n_k + 2$.
Remark: for $k \geqslant 4$ the inequality $n_k < 2^{k-1} - 2$ holds.
AMM 85, 1978, 825, E2667, J. R. Samborski; O. P. Lossers.

1.11 Answer: $(p - 1)/2(q - 1)$.
AMM 74, 1967, 601–602, 5392, H. Hahn; C. R. MacCluer.

1.12 Answer: f is continuous except for the points of the form $0.x_1 x_2 \ldots x_n 000 \ldots$ where f may not be left continuous.
  f is differentiable if $\sigma = Id$ or $\sigma(n) = 9 - n$. In the other case f is nowhere differentiable.
  f is Riemann integrable and $\int_0^1 f(x)\,dx = 1/2$.
AMM 87, 1980, 61–62, E2738, M. W. Ecker; A. Guzman.

1.13 [O] 437.

1.14 Hint: a more general fact holds: If for some $R \in \mathbb{N}$ $\lim_{k \to \infty} n_k^{1/R^k} = \infty$, then $\Sigma_{k=1}^{\infty} 1/n_k$ is not an algebraic number of degree less than R.
  To prove this the Liouville theorem may be used: If a is an algebraic number of degree R, then there exists $c = c(a) > 0$ such that the inequality $0 < |a - p/q| < c/q^R$ does not hold for any integers p and q.
[CHM] 6/1960, N.19.

1.15 *AMM 93*, 1986, 307–308, 6484, A. V. Nabutovskii.

1.16 *AMM 92*, 1985, 736–737, E2923, P. Erdös, C. Spiro; U. Abel.

1.17 Hint: remember the Euler identity: For $a \in \mathbb{C}$, $|a| > 1$, $\Pi_{n=1}^{\infty} (1 - a^{-n}) = \Sigma_{n=-\infty}^{\infty} (-1)^n a^{-(3n^2+n)/2}$.

AMM 87, 1980, 408–409, 6233, J. Lynch, J. Mycielski; P. A. Vojta, P. Bundschuh.

1.18   Hint: $-g(\theta) + g(1 - \theta) - (1 - \theta)^{-1} + \theta^{-1} = \pi \cot \pi\theta$. If $\theta \in \mathbb{Q}$ then $\cot \pi\theta$ is an algebraic number.
AMM 71, 1964, 103–104, 5072, W. E. Briggs; S. Heller.

1.19   *AMM 91*, 1984, 652–653, 6424, M. J. Dixon.

1.20   [K–N] 2.5.29.

1.21   [K–N] 2.5.30.

1.22   [K–N] 2.5.31, 32.

1.23   Answer: $n^2/2$ for even n, $(n^2 - 1)/2$ for odd n.

1.24   Hint: associate to each element of the sequence a pair of natural numbers: the lengths of the maximal decreasing and increasing subsequences beginning from this element. This correspondence is injective.
Erdös-Szekeres theorem, *J. London Math. Soc. 34*, 1959, 352, A. Seidenberg.

1.25   *EM 27*, 1972, 65, 647, P. Erdös; J. Fehér.

1.26   Answer: all but the powers of 2.

1.27   Hint: for irrational x remember the equipartition property of the sequence $(nx(\bmod 1))$.
Answer: in the first case the series converges only for integer x.

1.28   Hint: $\log_{10} 2$ is irrational.

1.29   [G] 219.

1.30   Hint: let (b_n) denote the sequence $(n/(n + 1))$ and let (a_n) contain the remaining rational numbers. Consider any sequence (c_n) such that $\sum_{n=1}^{\infty} c_n = \infty$. Then one can choose $k_1, k_2, \ldots$ so that $a_1 + a_1 b_1 + a_1 b_1 b_2 + \cdots + a_1 b_1 \cdots b_{k_1} \geqslant c_1$, $a_1 b_1 \cdots b_{k_1}(a_2 + a_2 b_{k_1+1} + \cdots + a_2 b_{k_1+1} \cdots b_{k_2}) \geqslant c_2$ etc.
AMM 67, 1960, 387–388, 4851, Z. A. Melzak; S. J. Einhorn.

1.31   Answer: if (x_n) has an accumulation point c, then $A \subseteq \{c, c/2, c/3, \ldots\}$. If $x_n \to \infty$, then $\lim_{n \to \infty} \exp(2\pi i x_n/a) = 1$. Denote $Y = \{a^{-1} : a \in A\}$, $Y_z = Y \cap (0, z)$. Then $\lim_{n \to \infty} \int_{Y_z} \exp(2\pi i x_n y) \, dy = m(Y_z)$, where m is the Lebesgue measure. The integrals converge to zero, hence $m(Y) = 0$.
AMM 71, 1964, 332–334, 5090, F. Suvorov; I. J. Schoenberg.

1.32   Answer: such a set exists.

AMM 75, 1968, 795–796, 5513, L. F. Meyers; J. C. Morgan II, O. P. Giesy.

1.33   Answer: for example the sequence $(13/3^n)$.
AMM 80, 1973, 815–817, 5851, D. Lind; D. Coppersmith.

1.34   Hint: $f_0(x) \pm f_1(x) = \exp(\pm x)$, hence $\exp(\pm(x_1 + \cdots + x_m)) = \Pi_{k=1}^{m}$ $(f_0 \pm f_1)(x_k)$.
AMM 74, 1967, 594, E1849, G. Purdy; M. F. Neuts.

1.35   Hint: $\mathbb{C}^* \simeq \mathbb{R} \times (\mathbb{R}/\mathbb{Z})$ so it suffices to show isomorphism $\mathbb{R} \times (\mathbb{R}/\mathbb{Z}) \simeq \mathbb{R}/\mathbb{Z}$.
AMM 90, 1983, 201–202, L. Richard Duffy.

**1.37   *AMM 75*, 1968, 918–919, 5538, L. F. Meyers; R. D. Berlin, L. J. Lardy.

**1.38   *Bull. Austral. Math. Soc. 34*, 1986, 473–475, S. A. Morris.

1.39   Hint: use the Cantor theorem on nested families of segments.
AMM 84, 1977, 827–828, E2613, D. E. Knuth; B. Gunter, O. P. Lossers; [B] Ex.10.2; *CMB 16*, 1973, 435–437, C. Eberhardt, J. B. Fugate, L. Mohler; *Coll. Math. 25*, 1972, 79–80, G. J. O. Jameson.

**1.40   *NAW 14*, 1966, 274–275, 71, K. A. Post.

1.41   Hint: each real number can be represented as $\Sigma_{n=1}^{\infty}\varepsilon_{2n-1}/(2n-1)$ and as $\Sigma_{n=1}^{\infty}\varepsilon_{2n}/(2n)$.

1.42   Hint: compare the multiplicities of the roots of the denominator and the numerator.

**1.43   *AMM 69*, 1962, 58–59, E1467, D. G. de Figueiredo; C. F. Pinzka, W. C. Waterhouse.

1.44   Hint: consider the coefficients of the polynomial divided by $x - 2$.
AMM 69, 1962, 568, E1493, L. Flatto, N. C. Hsu, A. G. Konheim; N. G. Gunderson.

1.45   Answer: one can assume that $Q(x) = x^n + a_{n-1}x^{n-1} + \cdots + a_0$ with integer a_j. Then there exists $k \in \mathbb{Z}$ such that $Q(k+1) - Q(k) > 1$. Consider an integer h such that $Q(k) < h < Q(k+1)$. There exists $r \in (k, k+1)$ such that $Q(r) = h$. r should be rational which is absurd.

**1.46   *AMM 72*, 1965, 550, E1696, D. I. A. Cohen, R. Greenberg; C. Chouteau, M. J. Pascual; see 7.58.

1.47   Hint: this polynomial is of the form $PQ/(P, Q)^2$.

1.48   Hint: show the following lemma: If b is a polynomial of degree at least 2, then for every m there exists at most one monic polynomial of degree m which commutes with b.

NAW 15, 1967, 160–161, 105, A. Menalda.

1.49   *AMM 70*, 1963, 671, E1540, A. Rosenfeld, M. S. Klamkin.

1.50   Answer: $(n - 1)/2$.
MS 3, 1976, 1702.

1.51   Hint: expand f/g into simple fractions, multiply by x and take the limit at infinity.

1.52   Answer: the sum vanishes if a_k are mutually different and is equal to ∞ otherwise.

1.53   Hint: the roots of the derivative satisfy the equation $(x - a_1)^{-1} + \cdots + (x - a_n)^{-1} = 0$.

1.54   Hint: $B_n(0) = B_n(1)$.
Answer: $B_n(x) = \Sigma_{k=0}^{n} \binom{n}{k} B_k x^{n-k}$, $S_n(m) = (B_{n+1}(m) - B_{n+1})/(n + 1)$.
[Po] Section 12.

1.55   Hint: $\Sigma_{k=1}^{n} a_k/k = \int_0^1 (\Sigma_{k=1}^{n} a_k x^{k-1}) \, dx$.
AMM 72, 1965, 318–319, E1682, V. R. Rao Uppuluri; J. Williams, P. A. Scheinok, W. J. Hartman.

1.57   Hint: remember the isoperimetric property of regular polygons.
AMM 71, 1964, 796, E1634, E. Just, N. Schaumberger; D. C. B. Marsh.

1.58   Hint: $t \log t \geqslant t - 1$.
e.g. *AMM 65*, 1958, 124, E1274, P. G. Kirmser, B. Marzetta.

1.59   Answer: this sum is maximal for the permutation reversing order of a_ks.

1.60   Hint: consider the polynomial $P(x) = (x - a_1) \cdots (x - a_n)$:
   (i) Differentiate P $n - 2$ times.
   (ii) Differentiate P $n - k - 1$ times, change the variable $x = 1/y$ and differentiate $k - 1$ times more.
   (iii) Prove by induction.
K, 4, 1980, M565.

1.61   *NAW 29*, 1981, 59–70, P. Noordzij; *AMM 89*, 1982, 757–758, 779–783, C. Kimberling; R. Breusch, T. Jager.

1.62   Answer: observe that the difference between the left and right hand sides is periodic of period $1/n$ and vanishes on $[0, 1/n)$.

1.63   Hint: one may restrict attention to the case $x \geqslant 1/n$, $x = p/q$, $q \leqslant n$, $(p, q) = 1$. Define natural numbers a_i, b_i, $i = 1, \ldots, n$, by $ip = a_i q + b_i$, $0 \leqslant b_i < n$. Then $[ix] = a_i$. Show that $\Sigma_{i=1}^{n} b_i/i \geqslant q - 1$.
See 1.59; *NAW 18*, 1970, 93–95, 215, O. P. Lossers.

1.64 *CMB 26*, 1983, 507, P. 329, M. Agimo; A. Facchini.

1.65 See for example [OD] 38.

1.66 **Hint:** consider the sequence $t_n = 2n - u_n$.
AMM 67, 1060, 380, E1382, I. Connell, A. Korsak.

1.67 **Hint:** $\sqrt{n+1} - \sqrt{n} \to 0$.
Answer: $\mathbb{R}$.
Remark: the answer is the same for any sequence (a_n) satisfying $a_n \to \infty$, $(a_n - a_{n-1}) \to 0$.

1.68 **Hint:** $\sum_{n=0}^{\infty} (a^n - b^n)z^n = \sum_{n=0}^{\infty} a^n z^n - \sum_{n=0}^{\infty} b^n z^n$ in its domain of convergence.
Answer: $\max(|a|, |b|)$.

1.69 *AMM 69*, 1962, 568–569, E1495, D. Hayden; J. Brown.

1.70 **Hint:** no regularity assumption on f is needed.
Remark: the function $f(x) = |x|^{2^{1/r}}$ satisfies the relation $f^r(x) = x^2$ but only for real x.
AMM 87, 1980, 252–263, R. E. Rice, B. Schweizer, A. Sklar.

1.72 **Answer:** $(Tf)(x) = f(x+1)$.

1.73 **Answer:** if $P - \lambda$ has a root of multiplicity m, then the derivative has a root of multiplicity $m - 1$. Hence the maximal multiplicity is $\leqslant 2$. The equality is satisfied for $\lambda = P(x)$ if $P'(x) = 0$.

1.74 **Hint:** let $z_1, \ldots, z_n$ be the roots of f, $|z_1|, \ldots, |z_k| > 1$, $|z_{k+1}|, \ldots, |z_n| \leqslant 1$, $f_1 = \Pi_{j=1}^{k}(z - z_j)$, $f_2 = \Pi_{j=k+1}^{n}(z - z_j)$, $g_1 = \Pi_{j=1}^{k}(z - 1/z_j)$, $g_2 = \Pi_{j=k+1}^{n}(z - 1/z_j)$, $g = g_1 g_2$, $A = f_1 g_2$, $B = f_2 g_1$. Compare the coefficients in fg and AB.
Remark: a_n can be arbitrary reals.
AMM 68, 1961, 387–388, 4909, D. J. Newman; R. Breusch.

1.76 **Hint:** λ is also a root of $(z - 1)f(z)$.

1.77 *AMM 71*, 1964, 322–323, E1598, J. Rainwater; A. Koler, J. E. Wilkins Jr.

1.78 **Hint:** compute the values at the roots of the derivative.

1.79 **Answer:** $n \leqslant 7$.
AMM 95, 1988, 140–141, E3078, Z. Rubinstein; W. O. Egerland.

1.80 **Hint:** add up the values of the polynomial at the nth roots of 1.
Answer: $M_n = 2$.

1.81 **Answer:** $\sqrt{3 - \sqrt{5}}$.
[CHM] 2/1956, F.2.

1.82 **Answer:** the minimum is attained for $z_1 = z_2 = \overline{z_3} = \overline{z_4}$, $|z_1| = r$.
AMM 84, 1977, 741, E2600, R. Evans; E. Johnston.

1.83 *NAW 2*, 1984, 697.

1.84 **Hint:** arrange $z_1, \ldots, z_n$, $-(z_1 + \cdots + z_n)$ in a sequence with increasing arguments. Compare the perimeter and the diameter of the obtained polygon.

1.85 **Hint:** for $f(x) = ((x+k)/(x+n+1))^x$ $(\log f)' = \int_{x+k}^{x+n+1} (x/\tau - 1)\tau^{-1} d\tau$.
EM 42, 1987, 173–174, 166, L. Kuipers.

1.87 **Hint:** $ab > 0$ is a necessary condition for the existence of solutions.
Answer: if $b > 0$ and $ab \neq 4$, then there exists a nontrivial solution. If $b < 0$, then the system has no solution or it has two or one (double) solutions according to the sign of the expression $q^{1/(q+1)} + q^{-q/(q+1)} - (ab)^{1/2}$, where $q = (a/b)^{1/2}$.
[L] 4.33.

1.88 **Answer:** $w \in \{x, y, z\}$, $x + y + z = w$.

1.90 **Answer:** $(x, y, z) = (3, 3, 3)$.
K, 2, 1978, 46.

1.91 **Answer:** for $c < 0$.

1.92 **Hint:** it suffices to show that the simplex $\{(x_1, \ldots, x_n) \in \mathbb{R}^n: 0 \leqslant x_1 \leqslant \cdots \leqslant x_n \leqslant 1\}$ is contained in the ball centered at $(1/(n+1), \ldots, n/(n+1))$ of radius $f(0)^{1/2}$.
Based on ideas from *NAW 4*, 1986, 162, 757, J. van de Lune.

1.93 **Answer:** $a_{n+9} = a_n$ for every n.
AMM 92, 1985, 218, 6439, M. Brown; J. F. Slifker.

1.94 *CMB 13*, 1970, 153–154, P. 161, H. S. M. Coxeter; S. Reich.

1.95 *CMB 26*, 1983, 254–255, P. 319, M. Bencze.

1.96 **Hint:** interpret φ as a positive definite function on the Cantor group.
Remark: if p is dyadic rational, then almost all a_γ can be chosen equal to zero.

1.97 *Proc. AMS 11*, 1960, 324–326, D. Wesler; *AMM 83*, 1976, 552–554, R. B. Crittenden, L. G. Swanson.

1.98 **Hint:** the smallest square satisfying this condition has the side of length $3/2$.
D 1987, 7; S. Mazur problem; [MV] 221; *J. Comb. Th. 5*, 1968, 126–134, A. Meir, L. Moser.

1.99 **Hint:** it suffices to show that the unit square can be covered by a finite number of these squares.
NAW 24, 1976, 104.

1.100 **Answer:** the sum of areas equals $\pi^5/90$.

1.101 *AMM 82*, 1975, 56–59, D. W. DeTemple.

1.102 **Answer:** $A_n \to \pi a^2/2$, $L_n/n \to 2a$.

1.103 **Answer:** the diameter equals $\sqrt{n} - 1$ so for $n \geq 10$ this ball is not contained in the cube!

1.104 **Answer:** the foci are on the line $y = 1/4$.

1.105 **Hint:** for every A, B and $\varepsilon > 0$ there exists a point C such that $d(A, C) < \varepsilon$ and the distances $d(A, C)$, $d(B, C)$ are rational.
AMM 64, 1957, 24, Putnam Competition 1956, I-6.

1.106 **Hint:** consider lattice points.
Proc. AMS 4, 1953, 810–815, F.-S. Beckman, D. A. Quarles; see *AMM 90*, 1983, 200, Th.M. Rassias, and for generalizations *Sib. Mat. Zh.* 27/1, 1986, 79–85, A. V. Kuz'minykh.

1.107 **Hint:** $|a - b| = 1$ if and only if there exist intervals I_1, I_2 in $\mathbb{Z}$ of length k such that $I_1 \cap I_2 = \{a, b\}$.

1.108 **Hint:** consider $a \neq b$ and the iterates $a_k = f^k(a)$, $b_k = f^k(b)$. For some $k = k(a)$ $d(a, a_k) < \varepsilon$ and $d(b, b_k) < \varepsilon$. Then show that $f(E)$ is dense in E and $d(f(a), f(b)) = d(a, b)$.
[Di] 3.17.4.

1.109 *AMM 70*, 1963, 95–96, E1515, J. Lambek, L. Moser, L. F. Meyers.

1.110 **Answer:** consider the line passing by $(x_0, y_0, 0)$ and contained in the plane perpendicular to the vector $(x_0, y_0, 0)$ which forms the angle $\pi/2 - \arctan(x_0^2 + y_0^2)^{1/2}$ with the plane $z = 0$.

1.111 **Answer:** e.g. $x = |G|^{-1} \Sigma_{g \in G} g(0)$.
Remark: the following Kakutani theorem is a deep generalization of this fact: If $\mathscr{F}$ is uniformly continuous group of affine transformations of a convex compact subset C of a locally convex topological vector space, then $\mathscr{F}$ has a common fixed point in C.

For a discussion on this theorem, its applications (for example, in the proof of existence of the Haar measure on compact groups) and generalizations (the Ryll-Nardzewski theorem) see for example J. Dugundji, A. Granas *Fixed Point Theory*, PWN, Warszawa, 1982, 75, 100–101, 181–182, and (Markov-Kakutani theorem on commuting families of transformations) [R-S] V.20.

1.112   **Answer:** $n(n-2)(2n-5)/24$ if n is even, $(n-1)(n-3)(2n-1)/24$ if n is odd.
AMM 66, 1959, 234–235, E1326, P. L. Chessin; C. S. Ogilvy.

1.113   **Answer:** the maximal area is $(3+\sqrt{3})/2$.
AMM 69, 1962, 314–315, E1484, J. H. Heinbochel, R. R. Korfhage; D. C. B. Marsh, M. Goldberg.

1.114   **Hint:** if P is a polygon with vertices in lattice points, then its area is equal to #(lattice points in the interior of P)+ #(lattice points on the boundary of P)/2 − 1.

1.115   **Hint:** for some integer N $\iint_P e^{2\pi i N(x+y)}\,dxdy=0$. Of course, there are also completely elementary solutions.

1.116   **Answer:** $a=2\pi, b=c=1$.

1.117   *AMM 71*, 1964, 637, Putnam Competition 1963, II-4.

1.118   **Answer:** the minimum is attained if B moves along the line perpendicular to AB and the trace of A consists of two segments and an arc. The minimal value is equal to $\sqrt{3}+\dfrac{\pi}{3}$.

1.119   **Answer:** $y=2^{-1/2}d(1-\cosh(2x/d)+2^{-1/2}\sinh(2x/d))$, where d is the side of the "wheel," $x\in[0,d]$.
AMM 72, 1965, 82–83, E1668, D. G. Wilson; D. C. B. Marsh.

1.120   **Hint:** if x_k denotes the wine concentration in the kth cup, then $dx_1/dt+x_1=1$, $dx_k/dt+x_k=x_{k-1}$, $k\geqslant2$.
Answer: the final concentration in the kth cup is $1-e^{-1}\Sigma_{j=1}^{k-1}1/j!$.
AMM 28, 1921, 143–145, 2791, C. Gilpin; R. E. Gaines.

1.121   *AMM 83*, 1976, 745–747, E2544, H. Cohn; G. Kermayer.

1.123   **Hint:** start from any point by car with a large amount of gasoline and take the gasoline from the passed stations. Look for the station where the fuel indicator shows the minimal value.

1.124   **Answer:** yes. It suffices that the torpedo boat moves along a spiral around the submersion point. Find the equation of this curve.

1.125   **Hint:** first solve the problem for the equilateral triangle and then apply an affine transformation.

1.127   *AMM 88*, 1981, 619, S-25, J. Mycielski; V. P. Snaith.

1.128   *AMM 84*, 1977, 491, 6158, M. J. Pelling.

1.129   *AMM 84*, 1977, 659, E2598, E. Just; P. L. Montgomery.

1.130 *Aequationes Math. 24*, 1982, 298, P. 215, Th.M. Rassias.

1.131 $a_{43} \in \mathbb{N}$, $a_{44} \notin \mathbb{N}$.
NAW 30, 1982, 115, 331–332, 636, F. Göbel; H. W. Lenstra Jr.

1.133 **Answer:** this probability is equal to $6/\pi^2$.
AMM 89, 1982, 339–340, 6300, J. Garfunkel, L. Takács; [Ch] VI. 6, T.22.

1.134 *AMM 70*, 1963, 1108, 5006, E. Ehrhart; H. G. Eggleston.

Answers to Chapter 2

Sequences

2.1 **Answer:** the first limit equals k^2, the second 1.

2.2 **Hint:** $\lim_{n \to \infty} a_{n+1}/a_n = e$.

2.4 **Hint:** $a_{n+1}/a_n = (1 - (n+1)^{-2})^{(n+1)^2}(1 + 1/n)^{n+1/2} =: e^{s_1 + s_2}$. Show that $s_1 + s_2 < 0$. The Stirling formula may be useful.
Answer: $\lim_{n \to \infty} a_n = \sqrt{2\pi/e}$.
AMM 80, 1973, 436–437, E2352, M. Sholander.

2.5 *AMM 87*, 1980, 832–833, 6252, I. Tomescu; R. Breusch.

2.6 **Answer:** $\lim_{n \to \infty} a_n = 0$.

2.7 **Hint:** $x_n^n \to x$ if and only if $n(x_n - 1) \to \log x$.
AMM 82, 1975, 401–402, E2434, G. O'Brien; J. Pitt.

2.8 **Hint:** it is easier to show the convergence of the sequence of logarithms. For rational x the problem is easy; for irrational x the equipartition theorem is useful.
Answer: $(1 + [1/x])^{1/x - [1/x]}[1/x]^{1 - 1/x + [1/x]}$.
[CHM] 7/1950/I, S.15.

2.10   The following generalization is given in *AMM 84*, 1977, 629–630, J. Fine:

$((x+a)/(x-b))^{1/k} = \Pi_{n=1}^{\infty}(1+1/x_n)$, where $x_1 = x > b \geqslant (k-1)/2$ and $x_{n+1} = (bx_n^k(x_n+a) + a(x_n+1)^k(x_n-b))/(x_n^k(x_n+a) - (x_n+1)^k(x_n-b))$.

2.13   *AMM 69*, 1962, 762, Putnam Competition 1961, I-3.

2.14   Answer: $\log 4 - 1$.
AMM 85, 1978, 29, Putnam Competition 1977, B-1.

2.15   Answer: $e^{1/2}$.
[L] 4.43.

2.17   [MS] 144/32.

2.18   Answer: $1/a$.
[MS] 151/41.

2.19   Answer: the limit is finite for $p > 1$ and infinite for $p \leqslant 1$.

2.21   Hint: consider first $s = 1$.
Answer: $1/(e^s - 1)$.
NAW 30, 1982, 334–337, 640, I. J. Schoenberg; J. H. van Lint, D. Dijkstra.

2.24   *MS, 4*, 1977.

2.25   Answer: $C \log n - \pi^2/6 + O(\log n/n)$.
[CHM] 7/1950/II, S.19.

2.26   Hint: observe that $s_n = \binom{2n}{n-1}^{-1} \sum_{k=0}^{n-1} \binom{2n}{k}$ and apply the Stirling formula.
AMM 90, 1983, 572, 6375, Chang Gengzhe.

2.27   [Kl1] 3.45.

2.28   [Kl1] 3.122.

2.29   Answer: 1 if (a_n) is a constant sequence, $2/e$ otherwise.

2.30   Hint: taking the logarithms and dividing by n the identity
$\Pi_{k=0}^{n}\binom{n}{k} = \Pi_{k=1}^{n}(n+1-k)^{n+1-2k} = \Pi_{k=1}^{n}\left(\frac{n+1-k}{n+1}\right)^{n+1-2k}$ one gets a
Riemann sum.
[Kl1] 3.124; [P-S] I.2.51.

2.31   Answer: $(a_1 + \cdots + a_k)/k$.

2.32   Hint: use the identities $\cotan 2x + \cosec 2x = \cotan x$, $(\cosec 2x \cotan x)^{1/2} = 2^{-1/2}\cosec x$ if $a_1 < a_2$ and similar identities for cotanh and cosech if $a_1 > a_2$.

Answer:

$$\lim_{n \to \infty} a_n = \begin{cases} \dfrac{(a_2^2 - a_1^2)^{1/2}}{\arccos(a_1/a_2)} & \text{if } a_1 < a_2 \\[3mm] \dfrac{(a_1^2 - a_2^2)^{1/2}}{\text{ar } \cosh(a_1/a_2)} & \text{if } a_1 > a_2 \end{cases}$$

AMM 64, 1957, 112–113, E1223, D. J. Newman; [Kn] 8 p. 41, 91 p. 228; [Bm] I.8.

2.33   Hint: $I(a, b) = I(\sqrt{ab}, (a + b)/2)$.
[L] 5.38.2.

2.34   Hint: if p and q are integers, then $|\pi - p/q| > 1/q^{42}$.
Answer: 1.
Indag. Math. 15, 1953, 30–42, K. Mahler.

2.36   Hint: use the substitution $y_k = x_k/k!$.
Answer: $x_n = n! \sum_{k=2}^{n} (-1)^k/k!$.

2.37   [K13] 1/79.

2.38   [K13] 1/77.

2.39   *AMM 74*, 1967, 773, Putnam Competition 1966, A-6.

2.40   Answer: $(((79 + 3\sqrt{249})/2)^{1/3} + ((79 - 3\sqrt{249})/2)^{1/3} - 1)/3$.
AMM 64, 1957, 46–47, E1219, D. A. Robinson; D. Freedman.

2.41   Hint: $a_n = a_1(1 - d^n)/(1 - d)$, where $d = 1 - a_1 - b_1$.
AMM 65, 1958, 449, E1294, G. A. Harris Jr; B. H. Bissinger.

2.42   Answer: 2.
See also [OD] 30.

2.43   Answer: the limit exists if $|ab| < 1$ and equals $x^{a(b+1)/(1-ab)}$.
AMM 72, 1965, 908–909, E1731, G. Mavrigian; V. H. Arakeri.

2.44   Answer: 1.

2.46   Hint: (i) The problem has an interesting geometric interpretation; (ii) Calculate the sum of cubes of the lengths of segments obtained from $(0, 1)$ by dividing with points $p_1, \ldots, p_n$.
Answer: 1/3. [K13] 1/62; *AMM 72*, 1965, 734, Putnam Competition 1964, I-3.

2.47   Answer: $(\sqrt{5} - 1)/2$.
[MS] 145/33.

2.48   Hint: if $m_i = \max(|a_{3i}|, |a_{3i+1}|, |a_{3i+2}|)$, then $m_n \leqslant m_0 \prod_{k=0}^{n} (1 + w_{3k+2})^{-1}$.

2.49 **Hint:** $1/x_{n+1} = 1/x_n + 1/n^{1+t} + O(1)/n^{2+2t}$.
NAW 18, 1970, 308–309, 245, M. L. J. Hautus.

2.50 *NAW 18*, 1970, 309–310, 246, M. L. J. Hautus; H. G. ter Mosche.

2.51 **Answer:** $\pi/2$.

2.52 **Hint:** substitute $u_n = \sin^2 x_n$.

2.53 **Hint:** useful transformation $c_n = \cos t_n$ or $c_n = \cosh t_n$.
Answer: $a_n \to c^2/2$, $b_n \to -(c \sin c)/6$, where $c = \arccos a_0$.
AMM 89, 1982, 275, E2835, M. Golomb; E. H. Gonnet.

2.54 *AMM 77*, 1970, 388–389, C. S. Ogilvy.

2.55 **Hint:** the set $D = \{z : |z| \leqslant 1, |1+z| \leqslant 1\}$ is invariant for $f(z) = z^2 + z$ and $G = \{z : |z| \leqslant 1, |1+1/z| \leqslant 1\} \supseteq D$ is invariant for $g(z) = (z+1/4)^{1/2} - 1/2$. Observe that $f \circ g = Id$.

2.56 **Hint:** a more general fact is true: If $g : \mathbb{R}_+^k \to \mathbb{R}$ is a function increasing with respect to every variable and $f(x) = g(x, x, \ldots, x) > x$ for $0 < x < a$, $f(x) < x$ for $x > a$ and the sequence (a_n) satisfies the recurrent relation $a_n = g(a_{n-1}, a_{n-2}, \ldots, a_{n-k})$, then $\lim_{n \to \infty} a_n = a$.

2.57 *AMM 94*, 1987, 789–793, E3036, F. Lazebnik, Y. Pilipenko. An extensive bibliography is given.

2.58 **Hint:** $\Sigma_{n=0}^\infty n!/(2n+1)!! = \pi/2$.
Answer: $\lim_{n \to \infty} a_n = 1$ for $x = \dfrac{\pi}{2} - 1$.

2.59 Some ideas are taken from [AKR] III.103; [Be] 10.3; see 9.93.

2.60 See for example [Si] I.92.

2.62 **Hint:** the sequence (y_n) increases, (x_n) decreases.
AMM 89, 1982, 504–505, 6304, I. J. Schoenberg; P. S. Bruckman.

2.63 **Hint:** first consider a_0, $a_1 > 1$. Find a differential equation satisfied by the function $\Sigma_{n=0}^\infty a_n z^n$.
Answer: $((9e^{-2} - 1)a_0 + (1 - 5e^{-2})a_1)/8$.
[CHM] 7/1958, S.10.

2.64 **Hint:** it can be shown that $a_n = 1 + \Sigma (r+s+t)!/r! \, s! \, t!$, where the sum is taken over all natural r, s, t such that $2^r 3^s 6^t \leqslant n$.
Remark: the recurrent formula in this problem has an interpretation in renewal theory, see [Fe] XI Section 1.
MM 58/1, 1185; for a generalization see *Pacific J. Math. 126*, 1987, 227–241, P. Erdös et al.

2.65   R. A. Adams, *Sobolev Spaces*, Academic Press, New York, 1975, 86–88.

2.66   **Hint:** $a_{n+1} = (1 + b_n^2)/2$, $b_{n+1} = a_n - a_n^2/2$ for $n \geqslant 2$.
Answer: $a_n \to 2 - 2^{1/2}$, $b_n \to 2^{1/2} - 1$.
AMM 64, 1957, 435–436, E1245, M. S. Klamkin; N. J. Fine.

2.67   **Hint:** observe that eventually $a_n < b_n < c_n$. Write a recurrent equation.
AMM 66, 1959, 921–923, 4828, M. S. Klamkin; C. H. Cunkle.

2.68   **Hint:** $c_{n+1}^{-1} = c_n^{-1} + (1 - c_n)^{-1}$.
[CHM] 5/1960, P.12; *AMM 93*, 1986, 739–740, E3034, D. Cox; O. P. Lossers.

2.69   *AMM 75*, 1968, 546–548, E1955, D. J. Newman; R. K. Meany, J. H. van Lint.

2.70   **Hint:** compute $\lim_{n \to \infty} (x_{n+1}^{-2} - x_n^{-2})$. Another method: solve the (formal) differential equation $dx_n/dn = \sin x_n \simeq x_n - x_n^3/6$.
See [P-S] I.1.173, 174.

2.71   **Answer:** $a_n = (2n)^{-1/2} - 3 \cdot 2^{-5/2} n^{-3/2} \log n + O(n^{-3/2})$.
AMM 74, 1967, 334–335, 5375, B. Volk; S. Spital, Yu Chang, a generalization *Bull. Soc. Math. France 47*, 1919, 161–271, P. Fatou.

2.73   **Hint:** first prove convergence of the sequence $u_n + A_{n-1}$, where $A_n = \Sigma_{k=1}^n a_k$.

2.74   **Remark:** this is the comparison of the Raabe and Cauchy convergence criteria for series.

2.76   *AMM 71*, 1964, 636, Putnam Competition 1963, I-4.

2.77   **Hint:** it suffices to show that $\lim \sup_{n \to \infty} (n(a_1 + a_n)/(n+1)a_n)^n \geqslant 1$.
Remark: more generally $\lim \sup_{n \to \infty} ((a_1 + a_{n+p})/a_n)^n \geqslant e^p$.
[P] V. 3, 4, 5, 6, 13.

2.78   *AMM 75*, 1968, 903–904, E1994, T. M. K. Davison; B. Margoliz.

2.79   **Hint:** express a_n as a function of b_n and t, then interpret a_n as a partial sum of a series.
AMM 73, 1966, 414–415, E1760, I. I. Kolodner, S. Spital.

2.80   *AMM 75*, 1968, 915–916, 5531, R. O. Davies; A. Meir.

2.82   For example [R3] 9.14.

2.83   **Hint:** prove by induction that $k_n > 2^{n-2} k_1$.
[CHM] 1/1950, S.1.

2.84   **Hint:** show first existence of a convergent subsequence of the sequence (a_n/n).

2.85 **Answer:** yes, the limit is equal to $\sup_n a_n/n$.

2.86 [CHM] 7/1956, S.7.

2.87 **Hint:** consider the sequence $(s_n^3 - s_{n-1}^3)$, where $s_n = \Sigma_{k=1}^n a_k^2$.
AMM 90, 1983, 573, 6376, I. J. Schoenberg; A. Meir.

2.88 **Hint:** compare the radii of convergence of the series $\Sigma a_n x^n$ and $\Sigma s_n x^n$.

2.89 *AMM 68*, 1961, 301–302, 4899, R. C. Buck; S. A. Andrea.

2.90 **Answer:** $a_n = Pb_n + Q + o(1)$ for some constants P and Q.

2.91 **Answer:** if a is rational, then $F(a,b) = F(a, b+\varepsilon)$ for sufficiently small $\varepsilon > 0$.

2.92 **Hint:** consider separately the factors with a_j/π rational and irrational. In the second case use the equipartition theorem.

2.96 **Hint:** choose in the set of limit points of (x_n) a countable dense set and construct the sequence (p_n) using 0 and 1.

2.97 *Duke Math. J. 5*, 1939, 740–752 H. G. Barone; *AMM 75*, 1968, 51 P. Schaefer; see 3.30.

2.98 *AMM 90*, 1983, 707, E2919, E. Johnston; J. Browkin.

2.99 [CHM] 6/1961, S.13.

2.100 **Remark:** this theorem is not true in the multidimensional case.
AMM 83, 1976, 273, B. P. Hillam; *AMM 87*, 1980, 748–749, D. R. Smart.

2.101 [H-S] 18.8, 18.9; see 3.96.

2.102 **Hint:** consider the function $x + \log(a - x)$.
AMM 64, 1957, 649, Putnam Competition 1956, I-6.

2.103 **Hint:** consider the cases $a_0 \in (-1, 2)$, $a_0 \in \{-1, 2\}$.
Uspekhi Mat. Nauk. 38:6, 1983, 162, V.

2.104 **Hint:** check that $a_{n+1} = t^2 a_n + t a_{n-1}$.
EM 40, 1985, 154, 932, J. C. Binz.

2.105 **Hint:** develop $(s + 1/s)^{2n+1}$.
Answer: (i) if $|x| \leqslant 2$, (ii) for $x \in \{-2, 0, 2\}$.

2.107 **Answer:** the sequence (f_n) converges uniformly to -1 on $[-1, -\varepsilon]$ and to 1 on $[\varepsilon, 1]$ for every $\varepsilon > 0$.
[Di] 8.7.4.

2.108 *AMM 70*, 1963, 762–763, E1546, N. V. Glick; N. J. Fine.

2.109   *Math. Intell. 7/1*, 1985, 49, 85-1, B. Reznick.

2.110   **Answer:** the sequence converges for preimages of 0 and 3/4 only. [CHM] 3/1951, S.20.

2.111   [L] 5.13.

2.114   N. Bourbaki, *Topologie générale*, Hermann, Paris, 1974, X, Section 4.9.

2.115   **Answer:** $e^{-1/2}$.

2.117   [K11] 3.33.

2.118   **Answer:** $x_k^n \to k$, $(x_n^{2n} - n) \to 1/2$.

2.119   **Hint:** consider separately the sequences (a_{2n}) and (a_{2n+1}).
Math. Intell. 9/3, 1987, 41–43, J. W. Grossman, A. J. E. M. Janssen, D. L. A. Tjaden.

2.120   [MS] 190, 2.

2.121   [MS] 130, 10.

2.122   **Answer:** the limit is equal to $\dfrac{1}{2}$.

MS, 3, 1978.

2.123   **Hint:** $((\sqrt{5}+1)/2)^n + ((1-\sqrt{5})/2)^n$ is an integer.

Answers to Chapter 3
Series

3.1 **Hint:** $1/(n-1) = \sum_{j=1}^{\infty} n^{-j}$.
[CHM] 5/1959, N. 18.

3.2 **Hint:** $a_{n+1} = 1 + a_1 \cdots a_n$, $\sum_{k=1}^{n} a_k = 1 - (a_{n+1} - 1)^{-1}$.

3.3 **Hint:** $S_n = f_n f_{n+1}$.
Answer: $(\sqrt{5} - 1)/2$.
AMM 65, 1958, 452, 4747, B. J. Boyer; D. C. B. Marsh.

3.4 **Hint:** write a recurrent formula for $S(2, 2k)$ and $S(2, 2k+1)$.
AMM 92, 1985, 513, 6450, J. O. Shallit.

3.5 *AMM 82*, 1975, 931–933, A. D. Wadhwa; see also R. Honsberger, *Mathematical Plums*, MAA, 1979, Ch. 3; *AMM 86*, 1979, 372–374, R. Baillie.

3.6 **Answer:** $0 < a < 91$.
Moreover, if $f(n)$ denotes the number of zeros in the expansion of n in the base b, then the series $\sum_{n=1}^{\infty} a^{f(n)}/n^p$ converges for $0 < a < b^p - b + 1$.

3.8 **Answer:** $\log \pi/2$.
MM 1970, 40, 231.

3.10 *AMM 95*, 1988, 664, E3174, A. Wilansky; W. Hensgen; see [G] 42.

3.11 Hint: this series may converge.

3.12 Answer: such a sequence exists if and only if $\Sigma_{n=1}^{\infty} \sqrt{a_n} < \infty$.

3.13 [Kn] Section 14, 78.

3.14 Hint: see 2.96.

3.15 *AMM 69*, 1962, 1014–1015, 4986, U. C. Gupta; R. H. C. Newton.

3.17 Hint: consider $f_i(x)$ = number of $j \geqslant i$ such that $\Sigma_{n=i}^{j} u_n \leqslant x$. Observe that $\lim_{n \to \infty} n/u_n = 0$.
AMM 93, 1986, 571–572, E3000, P. Erdös; J. B. Wilker, L. E. Mattics.

3.18 *NAW 30*, 1982, 197, 647, M. L. J. Hautus.

3.19 Answer: the optimal constant C_p is $1/(1 - p)$. To check this consider the series $\Sigma_{n=1}^{\infty} q^{n-1}(1 - q)$ for q close to 1.

3.20 Hint: another equivalent condition is
(iii) There exists a $\delta > 0$ and an increasing positive function f on $(0, \delta]$
    such that $F(x) \leqslant (f(x) - f(y))/(x - y)$ for all $0 < y < x \leqslant \delta$.
AMM 92, 1985, 347–349, P. Biler; see 6.74.

3.21 Hint: another equivalent condition is
(iii) There exists a $\delta > 0$ and a decreasing positive function f on $[\delta, \infty)$
    such that $F(x) \leqslant (f(y) - f(x))/(x - y)$ for all $\delta \leqslant y < x$.
AMM 84, 1977, 138–139, E2558, M. J. Pelling; *AMM 92*, 1985, 347–349, P. Biler; see *AMM 93*, 1986, 496–497, M. J. Pelling.

3.22 Hint: first prove the convergence of the series $\Sigma_{n=1}^{\infty} (1 - \lambda_n/ \lambda_{n+1})/f(\lambda_{n+1})$.

3.25 [Sp] 22.3.

3.27 Hint: show that one can choose signs in such a manner that $\max_{1 \leqslant j \leqslant m} |\Sigma_{k=j}^{n} \varepsilon_k z_k| < 6 \max_{1 \leqslant j \leqslant n} |z_j|$.
[CHM] 7/1957, F.9; *C. R. Acad. Sci. Paris 225*, 1947, 516–518, A. Dvoretzky, Ch. Hanani; *Trans. AMS 70*, 1951, 177–194, A. Dvoretzky, E. Calabi; for example *AMM 94*, 1987, 788–789, E3028.

3.28 Answer: no. Consider the permutation which exchanges $n(n+1)/2$ and $n(n+3)/2$.
AMM 80, 1973, 317–318, E2343, G. A. Heuer; D. Monk.

3.29 Hint: permute the terms of the series $\Sigma_{n=2}^{\infty} (-1)^n/(n \log n)$.
Answer: in case (ii) $\Sigma x_{\sigma(n)}$ may have another sum.

AMM 90, 1983, 548, Putnam Competition 1982, A-6; *AMM 94*, 1987, 687–688, E3092, R. Luttman; V. Pambuccian.

3.30 **Hint:** Fix k and consider $a_{kn} = b_n - kb_{kn}$. Apply 2.117.

3.31 **Remark:** these are all topological types of these sets.
Coll. Math. 55, 1988, 323–327, J. A. Guthrie, J. E. Nymann, see on the same subject: *Hokkaido Math J. 10*, 1981, special issue, 348–360, Th.3, S. Koshi, H.-C. Lai; *Rend. di Mat. appl. 3*, 2, 1983, 229–238, t.3c, Prop. 2, E. Barone, where the results are not complete.

3.32 **Answer:** $e^x \int_a^x e^{-t} f_0(t)\, dt$.

3.34 *MM 57/5*, 1201.

3.35 **Answer:** for $a = \sqrt{bc}$.

3.37 **Answer:** this series diverges.

3.39 **Hint:** one can choose natural p and q such that q is odd and $|\pi/2 - p/q| < 1/q^2$, so $|\sin^p p| > 1 - 1/q$.
AMM 65, 1958, 718–719, 4770, J. R. Holdsworth, J. R. Smart; J. B. Kelly.

3.42 **Hint:** the series of derivatives converges uniformly to a function g. If $f = \int_0^x g$ is bounded, then f would be almost periodic of zero mean. One can prove even the estimate $|\Sigma_{n=p+1}^\infty n^{-1} \sin(x/n)| \geqslant |x|/p$.
AMM 72, 1965, 559–560, 5203, D. J. Newman; M. D. Mavinkurve; *RMS 94*, 1983/84, 16.

3.43 *AMM 75*, 1968, 206–207, 5466, B. Volk; M. E. Muldoon.

3.45 **Hint:** a series $\Sigma_{k=1}^\infty b_k$ with $0 < b_k < 1$ converges if and only if the product $\Pi_{k=1}^\infty (1 - b_k)$ converges.
See 2.93.

3.49 [S-P] 221.

3.50 **Hint:** consider $K = L = \{n^2 : n \in \mathbb{N}\}$ and $K + L$, $\overline{K+L}$. Recall the Lagrange four squares theorem.

3.51 *AMM 72*, 1965, 735, Putnam Competition 1964, II-5; see a generalization in *CMB 14*, 1971, 277–278, P.87, P. Erdős; *D 1987, 3*.

3.53 **Hint:** write a differential equation satisfied by f.
Answer: $f(x) = d^{-1} x^{-a/d} e^{x/d} \int_0^x e^{-t/d} t^{a/d - 1}\, dt$.
[CHM] 1/1950/II, S.17.

3.54 **Answer:** this series converges for all $z \in \mathbb{C}$.
[CHM] 8/1952, F.8.

3.55   [D] 2739.

3.56   **Hint:** compare the functions $f(x) = \sum_{n=1}^{\infty} (-1)^{n-1}/n^x$, $g(x) = -f(x)$, $h(x) = f(x) - 1$.
[S-P] 474.

3.57   **Hint:** $\sum_{n=1}^{\infty} x^n/n^x = x \int_0^1 e^{xu \log(1/u)} \, du$.
Answer: the limit is equal to -1.

3.58   **Hint:** $2S(x) = 1 - \sum_{n \in \mathbb{Z}} (-1)^n (1 + n^2 x)^{-1/2}$. Using the Poisson summation formula (see 8.90) obtain $S(x) = 1/2 - 1/\pi \int_1^{\infty} (t^2 - 1)^{-1/2} \omega/\sinh \omega t \, dt$, $\omega = \pi x^{-1/2}$.
NAW 2, 1984, 179–180, 677, A. J. E. M. Janssen.

3.59   **Hint:** consider more generally $\prod_{n=1}^{\infty} (1 + z^{2^n})$.

3.60   **Answer:** z for $|z| < 1$ and 1 for $|z| > 1$.
[VLA] 817.

3.61   **Hint:** compute the difference of two consecutive terms.
Answer: e.

3.62   **Hint:** the Stirling formula may be useful.
AMM 87, 1980, 391, N. Pippenger.

3.64   **Hint:** one may use the Borel-Cantelli lemma.
AMM 85, 1978, 570–572, L. Carlitz; *AMM 85*, 1978, 363–364, F. Stern.

3.65   **Answer:** this series converges if $\sum_{k=1}^{n} a_k/(1 + a_k) < 1$ and its sum is equal to $(1 - \sum_{k=1}^{n} a_k/(1 + a_k))^{-1}$.

3.68   [G] 49.

3.71   **Hint:** give an explicit formula for a_n.
Answer: $1/4$.
EM 41, 1986, 130–131, 929, M. Vowe; P. Streckeisen.

3.72   An interesting solution is in *AMM 90*, 1983, 562–566, B. Ross.

3.73   **Hint:** $n^{1/n} - 1 \geqslant \log n/n$, $e - (1 + 1/n)^n \geqslant 1/2n$.
[D] 2612.

3.74   **Hint:** $f(x + 1) = e - xf(x)$.

3.75   **Hint:** the expression on the left hand side is equal to $(2\pi i)^{-1} \int_{|z|=1} (z - ae^z)^{-1} \, dz$. See a generalization in [P-S] I.3.124.

3.76   *AMM 90*, 1983, 54, E2982, R. Graham, D. E. Knuth.

3.77   **Answer:** for $0 \leqslant x < 1$ the sum is equal to $1 - 2x$.

3.78   **Answer:** $x^2 - x + 1$.

3.79   **Hint:** sum up 2^n terms.

3.81   **Hint:** replacing the sum by the integral $\int_0^\infty (1 - (1 - 2^{-t})^n)\, dt$ we get a better estimate: $\log n/\log 2 + O(1)$.
[G] 38; [New] 102.

3.82   **Hint:** $[k\sqrt{2}]$ should be carefully estimated.
See [K–N] Th.3.4; *AMM 85*, 1978, 207–208, 6105, H. D. Ruderman; D. Borwein.

3.83   [Kl2] 1.24.

3.89   **Hint:** expand in power series $f(x) = (1 + x)(1 + x^3)^{-1/2}$ and $g(x) = (1 - x + x^2)(1 + x^3)^{-1/2}$. Observe that $fg \equiv 1$.

3.90   **Answer:** this series converges for $a + b > 1$ and diverges otherwise.
[D] 2714.

3.91   **Hint:** $n - \log k_n \to \gamma$.
Answer: e.

3.92   See 3.26.

3.93   More generally [Si] Section 75, 21–25; *AMM 87*, 1980, 817–819, C. C. Cowey, K. R. Davidson, R. P. Kaufman.

3.94   **Hint:** estimate the integrals over $[0, c\sqrt{n}]$, $[c\sqrt{n}, \infty)$, where $c > 1$ is a fixed real number.

3.95   **Hint:** observe that $\binom{2n}{n}^{-1} = (2n + 1) \int_0^1 (1 - u)^n u^n\, du$. Write a differential equation satisfied by the sum of this series.
Answer: $(1 - x)^{-1}(1 + (x/(1 - x))^{1/2} \arcsin x^{1/2})$ if $0 \leqslant x < 1$, $(1 - x)^{-1}(1 - (-x/(1 - x))^{1/2} \operatorname{arsinh}(-x)^{1/2})$ if $-1 < x \leqslant 0$.

3.96   **Hint:** the product converges a.e. for every positive sequence (a_n) converging monotonely to zero.
Answer: $a_1 = 2$, $a_n = 1/\log n$, see 2.101.
[CHM] 2/1960, S.11.

3.97   **Hint:** consider the simple fraction expansion of $1/(n^2 - m^2)$.
[Di] 5.24.

3.98   **Answer:** $\log 2 - 1/2$.

3.99   **Hint:** $\sum_{n=1}^\infty \dfrac{1}{n} \sum_{m=1}^\infty \dfrac{1}{n+2}\left(\dfrac{1}{m} - \dfrac{1}{m+n+2}\right)$

$$= \frac{1}{2} \Sigma_{n=1}^{\infty} \left(\frac{1}{n} - \frac{1}{n+2} \right) \left(\left(1 - \frac{1}{n+3} \right) + \left(\frac{1}{2} - \frac{1}{n+4} \right) + \cdots \right).$$

Answer: 7/4.
AMM 86, 1979, 752, Putnam Competition 1978, B-2.

3.100 [Hi1] 5.3.1.

3.101 [Di] 5.2.6.

3.102 **Answer:** for example $b_n = n \log n$, take the subsequence with indexes $3^{k^2} + 1, \ldots, 3^{k^2} + 3^{k^2}$.
[CHM] 1/1951, A.2; see [S-P] 524.

3.103 **Hint:** $(B_0 + \cdots + B_n)/(n+1) = B_n - (b_1 + 2b_2 + \cdots + nb_n)/(n+1)$, where $B_k = b_0 + \cdots + b_k$.
[Hi1] 11.5.3.

3.104 e.g. [G] 477. See a generalization in the hint to 8.78. A converse theorem: *AMM 81*, 1974, 908–909, E2433, R. T. Smythe, A. Stenger.

3.105 *AMM 74*, 1967, 773, Putnam Competition 1966, B-3.

3.106 **Hint:** observe that $\Sigma_{n=2^{k-1}+1}^{2^k} (na_n - (n-1)a_{n-1})^{-1} \geqslant 2^k/4a_{2^k}$.
[CHM] 1/1959, N.17.

3.107 **Hint:** prove that $\Sigma_{k=1}^{2^n} a_k \leqslant (1+1)(1+1/2) \cdots (1+1/2^{n-1})a_1$.
[CHM] 6/1958, S.9.

3.108 **Answer:** take for example $b_1 = 2a_1$, $b_n = a_n S_n^{-1/2 - \varepsilon}$, where $S_n = \Sigma_{k=1}^{n} a_k^2$. Convergence of the series $\Sigma_{n=1}^{\infty} b_n^2$ is guaranteed by the Dini theorem.
[CHM] 8/1956, S.8.

3.109 [K12] 1.11, 12; [K11] 2.17.

3.110 **Hint:** apply the Cauchy inequality for $x_j = a_j^{-1/2}$, $y_j = j a_j^{1/2}$, $j = 1, \ldots, n$. Estimate the coefficients of a_j.
K, *12*, 1986, M999.

3.111 *J. London Math. Soc. 3*, 1928, 205–211, K. Knopp.

3.112 **Hint:** choose the sequence (c_n) such that $c_1 \cdots c_n = (n+1)^n$ and apply the inequality $(a_1 \cdots a_n)^{1/n} = (a_1 c_1 \cdots a_n c_n)^{1/n}/(n+1) \leqslant (a_1 c_1 + \cdots + a_n c_n)/(n(n+1))$. Another proof can be given using the results of 3.109 and 3.111.
[P] XVI.4; [New] 16; [K11] 2.18.

3.113 *NAW 3*, 1985, 124, 720, M. Bencze.

3.114 *AMM 94*, 1987, 684–685, G. Létac; O. P. Lossers.

3.115 **Hint:** $a_2 < a_3 + a_4 < a_4 + a_5 + a_9 + a_{16} < \cdots$. Show that the indexes do not repeat hence the rests of the series $\Sigma_{n=1}^{\infty} a_n$ are greater than a_2. [New] 87.

3.117 **Hint:** it is easy to show that the series converges uniformly in $\{z : \arg |z - z_0| \leqslant \delta\}$ for every $\delta \in (0, \pi/2)$. [Ch] X.4, Th.1.

3.118 **Hint:** if the sequence (b_n) converges to zero, then the series $\Sigma_{n=1}^{\infty} b_n / n^{1+\varepsilon}$ converges for every $\varepsilon > 0$.

3.119 **Hint:** consider the Taylor expansion of f at z_0.

3.120 **Hint:** the sums of Dirichlet series are analytic.

3.121 **Hint:** for $s > 1$ $\Sigma_{n=1}^{\infty} (-1)^{n+1} / n^s = (1 - 2^{1-s}) \Sigma_{n=1}^{\infty} 1/n^s$.

3.122 **Hint:** the characteristic function φ of a measure μ satisfies $|\varphi(t)| < 1$ for $t \neq 0$ if and only if μ is not concentrated on $r\mathbb{Z}$ for some real r. Then for $p \geqslant 1/2$ and arbitrary $c \in \mathbb{R}$ $pe^{ict} + (1 - p)\varphi(t) \neq 0$. Consider the measure $\mu = \Sigma_{n=2}^{\infty} 2(n(n + 1))^{-1} \delta_{-\log(n+1)}$.
NAW 25, 1977, 200–201, 454, J. van de Lune; F. W. Steutel.

3.123 **Hint:** compute $e^x + e^{\omega x} + \cdots + e^{\omega^{k-1} x}$, where ω is a primitive root of 1 of degree k.

3.124 **Hint:** check that the expression under the sign of limit equals $e^{ax} \int_0^{x(a+n)} e^{-t} t^{n-1} / (n - 1)! \, dt$ or $e^{ax} P\{X_1 + \cdots + X_n \leqslant x(a + n)\}$, where X_i are independent identically distributed random variables of mean 1 and of exponential distribution. Apply the Central Limit Theorem.
Answer: the limit equals e^{ax} if $x > 1$, 0 if $x < 1$ and $e^a/2$ if $x = 1$.
For the case $a = 0$ see *AMM 86*, 1979, 788–790, E2723, A. Moy; R. Breusch, M. Barr, O. Jørsboe.
See also modifications of the problem and different proofs: *J. Indian Math. Soc. 3*, 1911, 128, S. Ramanujan, Question 294; *J. London Math. Soc. 3*, 1928, 225–232, G. Szegö; *Proc. London Math. Soc. 29*, 1928–29, 293–308, G. N. Watson; *Proc. Edinburgh Math. Soc. 3*, 1932/33, 201–206, E. T. Copson; *Proc. Indian Acad. Sci. A, 11*, 1940, 376–378, F. C. Auluck; *Bull. AMS 55*, 1949, 396–401, T. T. Cheng; *J. Indian Math. Soc. 2*, 1960, 343–366, J. Karamata; *Proc. AMS 14*, 1963, 564–568, J. D. Buckholz; *Proc. AMS 16*, 1965, 248–252, L. Carlitz; *AMM 70*, 1963, 906–907, 5054, R. Sinkhorn; A. G. Konheim; [P-S] II.211; [S-P] 87; [New] 96; [Kl1] 4.53; *CMB 3*, 1960, 297–298, P. 29, E. L. Whitney.

3.125 A full discussion of this Rayleigh formula is in *AMM 93*, 1986, 660–664, 6488, R. M. Young; R. William Gosper, E. Salamin, K. Zacharias.

3.126 **Answer:** one may construct a function f for which h is a.e. finite. In this way an example of the set in $\mathbb{R}^+$ of infinite measure whose intersection with the sequence (nc) is finite for almost all c is given, see *Mathematika 22*, 1975, 195–201, J. A. Haight.
Moreover for every $0 < a < b \int_a^b h(c)\,dc = \infty$ though $h < \infty$ a.e.
Z. Magyar, *On a conjecture about functions not belonging to L^2*, preprint *Math. Inst. Hungarian Acad. Sci. 4*, 1987; *Studia Sci Math. Hung. 23*, 1989, 245–250, M Horváth; see 6.25, 6.65, 4.128.

3.127 [S-P] 553.

3.128 [K12] 4.4.

3.129 **Answer:** this product converges for $|x| < 1$ and arbitrary p and q and for $x = \pm 1$, $p > 1$, $q > 1/2$.
[D] 3083.

Answers to Chapter 4

Functions of One Real Variable

4.1 Hint: consider the factors indexed by k and $n-k+1$ together.
MM 58/1, 1208.

4.2 Hint: consider the auxiliary function $f(x)=\log(1/x-1)$ and apply the Jensen inequality.
MS, 5, 1977.

4.3 Hint: Observe that $\bigcup_j \bigcap_i A_{ij} \subseteq \bigcap_i \bigcup_j A_{ij}$. Give a probabilistic interpretation.

4.4 Hint: these inequalities follow from, for example, the weighted inequality $\Pi a_j^{w_j} \leqslant (\Sigma w_j a_j / \Sigma w_j)^{\Sigma w_j}$ with a suitable choice of a_j and w_j.
AMM 86, 1979, 224–225, E2691, Ž. M. Mijalković, J. B. Keller; R. Lyons.

4.5 Hint: use concavity of the logarithmic function and prove by induction.
AMM 94, 1987, 77–78, E3941, J. M. Borden, D. Mason; R. L. Young, O. P. Lossers.

4.6 Hint: consider auxiliary functions $\log(x^x/\Gamma(1+x))$, $\log(\Gamma(1+x)/(1+x)^{1+x})$.
AMM 89, 1982, 216–217, 6294, M. S. Klamkin.

4.7 **Hint:** the function $\log(\sin x/x)$ is concave.

4.8 **Hint:** one may assume that $1 = x < y$ (due to homogeneity). Then the inequality $M_1 < M_2$ is equivalent to $((z + 1/z)/2)^d \geqslant (z^a + 1/z^a)/2$, where $d = 1/r + 1/s$, $a = 1/s - 1/r$. Put $z = e^t$.
Answer: $M_1 < M_2$ if $(r - s)^2 \leqslant rs(r + s)$.
AMM 91, 1984, 261–262, 6403, J. L. Brenner; L. E. Mattics; see *AMM 93*, 1986, 742–743, E3040, J. L. Brenner; W. A. Newcomb, S. Philipp.

4.9 **Hint:** show that for $t > 1$ $\frac{3}{8}\log t - (t^3 - 1)/(t + 1)^3 > 0$ and put $t = (xy)^{1/3}$.
AMM 81, 1974, 879–883, T. P. Lin; *AMM 87*, 1980, 545–548, K. B. Stolarsky; *AMM 92*, 1985, 99–104, A. O. Pittenger. See also *AMM 95*, 1988, 262–264, E3142, Zhang Zaiming; R. Perez Marco.
Remark: a simple proof of the inequalities between harmonic, geometric, logarithmic, arithmetic and square means follows from the (obvious) inequalities $\sinh t/(\sinh^2 t + \cosh^2 t)^{1/2} < \tanh t < t < \sinh t < 2^{-1}\sinh 2t$ for $t = \log(y/x)^{1/2} > 0$.
AMM 92, 1985, 50, F. Burk.

4.10 **Hint:** for $n = 1$, $f(a, b)g(a, b) = a^2 b^2$. For $n = 2$, f must be homogeneous: $f(ta, tb) = t^2 f(a, b)$. Then prove that $F(a) = f(a, 1)$ is nonincreasing and $F(a)/a^2$ nondecreasing.
AMM 76, 1969, 98–100, 5563, D. E. Daykin, C. J. Eliezer; L. Carlitz.

4.11 **Hint:** prove by induction the inequality $\exp(t_1)((1 - t_1) + e_{n-1}) \leqslant \exp e_{n-1} = e_n$.
AMM 89, 1982, 453, Putnam Competition 1960, AM3, H. Flanders.

4.12 **Hint:** one may assume that $0 < x_j < 1$, $j = 1, \ldots, n$, and relabel x_j cyclically to get $x_3 = \min x_j$. Then $x_1^{x_2} + x_2^{x_3} \geqslant x_3^{x_2} + x_2^{x_3} \geqslant 1$.
Remark: this inequality cannot be improved. For $n = 4$, taking $x_1 = r^{-r^r}$, $x_2 = r^{-r}$, $x_3 = 1/r$, $x_4 = 1$ one has $3/r + 1 \geqslant 1$, so take $r \to \infty$.
A slightly generalized inequality

$$x_1^{x_2} + \cdots + x_n^{x_1} > 1 + (n - 2)\min(x_1^{x_2}, \ldots, x_n^{x_1})$$

AMM 87, 1980, 302–303, S6, M. S. Klamkin, A. Meir; D. Hammer.

4.13 **Hint:** the substitution of x_i and x_j by their arithmetic mean increases the expression on the left hand side.

4.14 *AMM 94*, 1987, 384–385, E3087, Weixuan Li, E. T. H. Wang; M. S. Klamkin, A. Meir; see also *Aequationes Math. 30*, 1986, 208–211, M. B. Subrahmanyam.

4.15 **Answer:** the equality holds for $x_k = (\tan((2k - 1)\pi/2n))^{-1}$, $k = 1, \ldots, n$.

4.16 **Hint:** it suffices to consider the case $0 < a_k < 1$.
AMM 85, 1978, 766–767, E2656, G. Tsintsifas; A. Stenger.

4.19 **Answer:** the equality holds if and only if $p = 1$, $x = k + 1/2$ with natural k.
NAW 2, 1984, 161, 690, M. J. Bastiaans, J. Boersma.

4.20 **Hint:** divide the interval $[0, 1]$ into $2n$ intervals of equal length. Choose x in an interval which does not contain any p_j.
Remark: this estimate can be improved.
AMM 87, 1980, 636, Putnam Competition 1979, A-6.

4.21 *AMM 87*, 1980, 761–762, 6247, M. Bencze.

4.22 **Hint:** it suffices to integrate the inequalities from the hypotheses.
[MS] 166, 6.

4.23 **Answer:** 1 for $0 < a < 1$, a for $a > 1$.

4.24 **Answer:** $-1/2e$.
[L] 4.40.4.

4.25 **Answer:** the limit does not exist. Denote by $f(x)$ the expression in question and suppose that L is the limit of $f(x)$. Then $L^2 = \sqrt{2}$ as $f(x)f(x^2) = \sqrt{1 + x}$. However for the recurrent sequence $x_1 = 0.98$, $x_{n+1} = x_n^{1/4}\, 2^{1/4} < f(x_1) < f(_2) < \cdots$.
AMM 68, 1961, 303–304, 4905, B. Gordon; P. Erdös, J. H. Roberts.

4.26 **Answer:** $p!$.

4.27 **Hint:** $S_n = \int_0^1 (1 - y^n)/(1 - y)\, dy$.
Answer: $g(x) = \int_0^1 (f(x) - f(xy))/(1 - y)\, dy$.
AMM 69, 1962, 239–240, 4946, M. S. Klamkin; W. H. M. Kantor, J. W. Wrench.

4.28 **Answer:** only for $a = e$.
[MS] 121, 1.

4.29 *EM 40*, 1985, 78, 925, P. Ivády.

4.30 **Hint:** for $y = 1 - e^{-x}$, $-\log(1 - y^a) < y^a \sum_{n=0}^{\infty} y^n/(n+1) < y^a (\sum_{n=0}^{\infty} y^n/(n+1))^a = x^a$.
AMM 86, 1979, 309, E2965, E. Beller; O. G. Ruehr.

4.31 *AMM 94*, 1987, 196–197, 6495, R. E. Shafer; W. A. Newcomb.

4.32 *AMM 92*, 1985, 516, 6453, R. Askey.

4.33 *AMM 74*, 1967, 1012–1013, E1902, D. Shelupsky; D. C. B. Marsh.

4.34 *EM 40*, 1985, 154, 934, P. Ivády.

4.35 *EM 39*, 1984, 103, 911, V. D. Mascioni.

4.37 **Hint:** a more general fact can be proved: if the functions f and g satisfy $f(0) = g(0) = 0$, $f(a) \leqslant a = g(b)$, $g(f(a)) = f(a)$, $f(x) < x$ for $0 < x < a$, $g(x) < x$ for $0 < x < f(a)$, f' decreases, g' increases, then $g \circ f(x) < f \circ g(x)$ for $0 < x < b$.
The inequality is true for $0 < x < x_0 \approx 1.3644$.
AMM 86, 1979, 706–707, E2720, R. P. Boas; P. Henrici.

4.40 *NAW 3*, 1985, 312, 734, 735, J. van de Lune.

4.43 **Answer:** this equation may have 0, 1, 2, or 3 roots.
K 1980, *1*, 48–50; *D* 1983, *6*.

4.44 **Hint:** put $t = \log x$ and study the equivalent equation $1/k = \log t/t$.

4.45 [MS] 129, 1.

4.46 *AMM 73*, 1966, 1127–1128, 5341, G. Purdy; M. Schulz, H. Pittie.

4.47 **Hint:** substitute $u = bx + a(1 - x)$.
Answer: $(b^b/a^a)^{1/(b-a)}/e$.
AMM 87, 1980, 636, Putnam Competition 1979, B-2.

4.48 **Hint:** the Wallis formula may be useful.
Answer: $\log(4/\pi)$.
AMM 91, 1984, 491, Putnam Competition 1983, B-5.

4.51 **Answer:** $\dfrac{\pi}{2} \log \pi$.

4.53 **Answer:** π.

4.54 **Answer:** 1.
[MS] 130, 9.

4.55 *NAW 15*, 1967, 172–173, 134, P. R. Masani; I. J. Schoenberg, F. W. Steutel.

4.56 **Hint:** $(e^x + 1)^{-1} + (e^{-x} + 1)^{-1} = 1$.
Answer: $\pi/4$.
[MS] 120, 4.

4.57 **Answer:** $w = \{z : |z| < 2/\pi\} \cup \{2i/\pi\} \cup \{-2i/\pi\}$.
NAW 14, 1966, 285–286, 108, A. Wilansky; P. van der Steen.

4.61 **Hint:** use the beta function.

4.63 **Answer:** 0.

4.64 **Hint:** compare the integrals with

$$\sum_{k=1}^{n} \int_{(k-1)\pi/n}^{k\pi/n} \frac{\sin k\pi/n}{1+\cos^2 nx}\,dx$$

The Fejér theorem may be useful, see 4.126, 8.19.

Answer: $\sqrt{2}$.
[CHM] 2/1949, R.1.

4.65 *AMM 91*, 1984, 652, 6482, G. Vrancken; *AMM 93*, 1986, 574; *AMM 93*, 1986, 488–489, D. R. Smith; W. Whitt.

4.67 **Hint:** show successively that $\int_0^\infty t^{-1}e^{-x-t}\,dt \approx 0$, p.v. $\int_{-x}^x t^{-1}e^{-x-t}\,dt$ $=e^{-x}\int_0^x t^{-1}(e^{-t}-e^t)\,dt$, $\int_0^x t^{-1}(e^{-t}-e^t)\,dt \approx \int_1^x t^{-1}e^t\,dt + O(1)$. Integrate by parts and observe that $\lim_{x\to\infty} x^n e^{-x}\int_1^x t^{-n-1}e^t\,dt = 0$.
[VLA] 781.

4.68 **Hint:** consider for example the integral $\int_0^a \int_0^a e^{-(x^2+y^2)}\,dxdy$.
[L] 6.20.

4.69 **Hint:** the condition from the assumption reads $\int_0^\infty f(x) \times \exp(-2(s-x/4)^2)\exp(x^2/8)\,dx \leqslant \exp(-s^2)$. Integrate with respect to s and use the Fubini theorem.

A generalization: if $\int_0^\infty f(x)e^{sx}\,dx \leqslant \exp(s^m)$ for some $m>1$ and $s>0$, then $\int_0^\infty f(x)\exp(\varepsilon x^{m/(m-1)})\,dx < \infty$ for an $\varepsilon>0$.

4.70 **Answer:** for example $\exp(-n\varepsilon^2/2)$.
Remark: this estimate is not optimal. Give a geometric interpretation of this inequality.

4.71 **Answer:** $b=\pi(\pi-2)/2(4-\pi)$, $c=\pi/(\pi-4)$, $d=-\pi^2/2(\pi-4)$.
AMM 71, 1964, 437–438, E1608, G. A. Heuer, D. Knudson; C. Comstock.

4.72 **Hint:** prove by induction or use analytic functions,

$$f(x) = -i \int_{1-i\infty}^{1+i\infty} \left(\frac{e^z - e^{-z}}{2z}\right)^{2k} e^{2xz}\,dz.$$

[CHM] 10/1958, R.7.

4.76 **Answer:** $\lim_{n\to\infty} u_n = 1/2$.

4.77 *NAW 15*, 1967, 73–74, T. van Ardenne-Ehrenfest, C. Visser; I. J. Schoenberg.

4.78 *AMM 92*, 1985, 58, E3072, L. Funar.

4.79 **Hint:** the assumption that P_m is a polynomial of the lowest degree with a nonreal root implies that P_{m-1} has a nonreal root.

AMM 87, 1980, 304–305, E2737, R. R. Wilson; A. J. Douglas, G. T. Vickers, F. B. Strauss.

4.80 **Answer:** no, as it is an almost periodic function of zero mean.
AMM 64, 1957, 50, 4602, C. M. Ablow, D. L. Johnson; G. Lorentz; see 8.82.

4.81 *Uspekhi Mat. Nauk 38*:6, 1983, 162; [MS] 188.

4.83 **Hint:** if c would be the only zero of f, then $f(x)\sin(x-c)$ would have a constant sign.

4.84 **Hint:** use the Baire category theorem.
[G] 56.

4.85 **Answer:** not necessarily.
[MS] 134, 17; *D* 1982, *11*.

4.87 *AMM 93*, 1986, 734–736, E3001, M. Slater, J. Propp, M. Bowron.

4.88 *AMM 68*, 1961, 562–563, N. McShane.

4.89 **Hint:** if f is a bijection with the Darboux property, then f is strictly monotone.
Answer: $f(x) = -x$.
AMM 85, 1978, 690, 6132, M. Eşanu.

4.91 **Hint:** consider for example the set $K = \{(r,t): 1/2 \leqslant r \leqslant 1\}$ and the mapping $f: K \to K$ defined by $f(r,t) = (1/(2-r), t+1-r)$.
AMM 80, 1973, 87, 5820, J. Cano; S. J. Bernau, R. D. Davies.

4.93 **Remark:** the interval $[a,b]$ can be replaced by a compact subset of a Banach space.
AMM 81, 1974, 506–507, D. F. Bailey.

4.94 *AMM 85*, 1978, 771, 6133, J. Cano, A. Grubler; R. B. Israel.

4.96 **Hint:** consider a such that for some pair p and q of rationals $f(a) \leqslant p < q \leqslant \lim_{x \to a} f(x)$ and analogously in the case $f(a) > \lim_{x \to a} f(x)$.
[Di] 13.3.3; [Sp] 21.24.

4.98 **Answer:** (ii) for example $f_n(x) = \sin(x - r_n)^{-1}$ for $x < r_n$, $f_n(x) = 0$ for $x \geqslant r_n$, where (r_n) is an enumeration of rationals and $f(x) = \sum_{n=1}^{\infty} f_n(x)/2^n$.
AMM 91, 1984, 445–446, 6412, F. S. Cater; P. I. Pastro.

4.99 **Hint:** observe that $f(x)/x$ is strictly decreasing and $f(x)/(1-x)$ is strictly increasing on $(0,1)$.
Remark: this is a generalization of the inequality $|\sin nx| \leqslant n|\sin x|$.
[CHM] 8/1958, R.6.

4.100 **Answer:** not always. For example, this limit does not exist for $f(n) = n$ if the distance from n to the set of powers of 2 is $\leqslant k$ and $f(n) = 0$ elsewhere. Of course if $k = \infty$ this limit exists.
AMM 91, 1984, 651, E2954, A. Meir; V. Grinberg; see *AMM 88*, 1981, 766, E2841, J. M. Steele; J. Muldowney.

4.102 *AMM 88*, 1981, 354–355, 6273, K. L. Chung; F. S. Cater; see *Proc. AMS 79*, 1980, 107–109, J. M. Steele.

4.104 *AMM 67*, 1960, 869, D. C. Benson.

4.106 A variation on *MM 57/5*, 1178.

4.107 [MS] 134, 16.

4.110 *AMM 75*, 1968, 44–45, D. Paine.

4.111 [Si] Section 114; [Kl2] 4.41.

4.112 **Hint:** use the functions g_n defined by $g_n(t) = 0$ for $t \leqslant 0$ or $t \geqslant 2/n$, $g_n(1/n) = 1$, and g_n is linear on $[0, 1/n]$, $[1/n, 2/n]$. Then take $f_n(t) = \Sigma_{k=0}^{\infty} g_n(t - r_k)/2^k$, where (r_k) is an enumeration of the rationals. [Di] 7.2.2.

4.113 **Answer:** for example $a_n = n!$, $K = \bigcup_{n=1}^{\infty} \{1/n\} \cup \{0\}$.

4.114 **Answer:** no. Consider for example the function $f_n(x) = (1 - d(x, [0, 1]\setminus U))^n$, where U is an open set of measure $1/2$ covering the rationals. *AMM 80*, 1973, 1062–1064, E2381, E. S. Langford; N. Wilson, H. Kestelman.

4.115 [L] 5.11.

4.116 **Hint:** approximate $x^{-n}f(x)$ by polynomials on $[c_n, 1]$, where c_n is such that $|f(x)| < 1/n$ on $[0, c_n]$.

4.118 **Hint:** use the Bernstein polynomials.

4.119 [G] 50.

4.120 [G] 54, see also 8.79.
Remark: the first example of such a universal power series was given by Pál and Fekete.

4.121 [P–S] I.1.96.

4.123 **Hint:** consider $g(x) = (f(xr) - f(r))/(x - 1)$, observe that $\lim_{x \to 1} g(x) = rf'(r)$.
AMM 84, 1977, 301, 6038, O. Strauch, J. Denmead Smith.

4.124 *NAW 15*, 1967, 168–169, 130, L. Kuipers.

4.125 *AMM 85*, 1978, 814–817, P. J. O'Hara, R. S. Rodriguez. Classical results: [P-S] I.2.147–194 and of course [K-N].

4.126 For example *AMM 75*, 1968, 735, Putnam Competition 1967, B-3; [CHM] 2/1949, R.1; [MS] 126, 2; [MS] 180, see also 4.64, 8.19.

4.127 **Answer:** for $a = 1$ and $g(s) = e^{-s}$ the function f is bounded.

4.128 **Hint:** let $G(x) = \Sigma_{n=1}^{\infty} f(nx)$. Show that G is lower semicontinuous and, then, that any set $\{x : G(x) \leqslant N\}$ cannot contain an open interval. Hence $\{x : G(x) = \infty\}$ is dense.
AMM 64, 1957, 119–120, 4670, K. L. Chung; A. R. Hyde, N. J. Fine.

4.129 **Hint:** consider the expression

$$\int_0^x (f(t))^2\, dt \cdot \int_0^x t f(t)\, dt - \int_0^x t(f(t))^2\, dt \cdot \int_0^x f(t)\, dt.$$

AMM 89, 1982, 451, Putnam Competition 1957, AM3, H. Flanders.

4.130 **Answer:** $f(1)/g(1)$.
[G] 61; [MS] 166, 3.

4.131 **Hint:** a useful substitution $g = f''$, $s = tu$ for $u > 0$ and $u = -v$, $s = t + v$ for $u < 0$.
Corollary: for $u = 1$ the considered expression increases to ess $\sup_{[a,b]} f$.
AMM 87, 1980, 670–671, S.15, J. L. Brenner; O. P. Lossers.

4.132 **Remark:** one can prove even $\int_{r-s}^r h > \int_{r-s}^r g$ and $\int_r^{r+s} h > \int_r^{r+s} g$.
AMM 89, 1982, 428, E2867, D. K. Mick; D. M. Wells, B. L. Montgomery.

4.133 **Answer:** if F and G are the primitives of f and g, $F(0) = G(0) = 0$, then the expression on right hand side is the squared length of the curve $(F(t), G(t))$, $0 \leqslant t \leqslant 1$; on left hand side it is the square of the length of segment $[(F(0), G(0)), (F(1), G(1))]$.
[MS] 155, 48.

4.134 For example [G] 107.

4.135 *Monatsh. Math. 69*, 1965, 362–367, D. Willett, J. S. W. Wong.

4.137 **Hint:** consider the function $P(s, t) = \int_0^t u^s p(u)\, du$, $t > 0$.
Remark: the continuity assumption can be weakened.
NAW 2, 1984, 162, 696, J. van de Lune.

4.138 *AMM 73*, 1966, 212–213, 5266, J. D. Pryde.

4.139 **Answer:** 1/4 and $s(f) = 1/4$ if and only if $f \equiv 1/2$.
AMM 86, 1979, 508, E2707, L. Shapiro; T. Jager.

4.140   **Answer:** yes, $F(x) = x^2 \cos(1/x) - 2 \int_0^x t \cos(1/t)\, dt$.

4.141   *AMM 94*, 1987, 82–83, 6492, A. Beck; O. P. Lossers.

4.142   **Hint:** consider the function fe^g.

4.143   **Answer:** yes, for example $f(x) = \Sigma_{n=1}^{\infty} g(n!x)/(n!)^2$, where g is 1-periodic and $g(x) = x(1 - 4x^2)$ for $|x| \leqslant 1/2$.
AMM 82, 1975, 415–416, 5955, F. D. Hammer; W. Knight; see also 5.93.

4.144   *MM 58/2*, Q696.

4.145   **Answer:** no, if f' has a strict local extremum at x; yes if f' is monotone.
Remark: some additional information on this Pompeiu problem is in *Revue Roumaine Math. Pures et Appl. 27*, 1982, 1053–1058, V. Nistor. See *CMB 14*, 1971, 603–604, P.59, R. Dowling; R. Park.

4.146   *AMM 93*, 1986, 733, E3173, H. W. Oliver.

4.147   *AMM 93*, 1986, 738–739, E3033, M. McAsey, L. A. Rubel; V. D. Mascioni, R. Spigler.

4.148   *NAW 21*, 1973, 288–290, 377, N. G. de Bruijn; F. W. Steutel, A. A. Jagers.

4.149   *AMM 68*, 1961, 365, J. P. Evans.

4.152   **Hint:** if the subspace of $C[0,1]$ spanned by (f_n) is contained in $C^1[0,1]$, then the operator d/dx would be closed.
AMM 70, 1963, 340, 5009, A. Wilansky; H. A. Gindler.

4.153   [CHM] 8/1961, R.10.

4.154   *AMM 91*, 1984, 314–315, 6293, W. M. Kahan.

4.155   *EM 38*, 1983, 891, 24, H.-J. Seiffert.

4.158   **Hint:** $(\int_a^b f)^2 - \int_a^b f^3 = 2 \int_a^b \int_a^y f(y)f(v)[1 - f'(v)]\, dv dy$.

4.159   *EM 37*, 1982, 62, 877, H.-J. Seiffert.

4.160   *AMM 80*, 1973, 755–760, O. Shisha.

4.161   **Hint:** integrate by parts $\int gf''$, where $g(x) = x_+ - 2(x - 1)_+ + (x - 2)_+$.
NAW 13, 1965, 241–242, 36, T. van Ardenne-Ehrenfest; I. J. Schoenberg.

4.162   **Hint:** apply the Rolle theorem twice to the function $(t-a)(t-b)f(x)-(x-a)(x-b)f(t)$ with fixed $a<x<b$ and get $|f(x)| \leqslant M(x-a)(b-x)/2$.

4.163   [MS] 190, 2, IV.

4.165 [MS] 135, 19.

4.166 **Hint:** consider the function f on an interval $[c, d] \subseteq (a, b)$ and add up a quadratic function.
AMM 68, 1961, 673–674, 4906, P. L. Butzer; J. Barlaz.

4.168 **Hint:** $\int_0^1 (x - 1/2)^n f(x) \, dx = 1$.
[Kl2] 3.56.

4.169 **Hint:** compare f with a function of the form $c \sin(x - a)$, c sufficiently small.
See *CMB 24*, 1981, 255–256, P.283, F. S. Cater.

4.170 *AMM 85*, 1978, 28, Putnam Competition 1976, A-6.

4.173 **Hint:** apply the Cauchy mean value theorem to the function $e^x f(x)$.
For generalizations of this classical Hardy's result see *AMM 81*, 1974, 92–93, 5875, Anon; W. C. Waterhouse; *AMM 85*, 1978, 685–686, E2663, M. Solomon, H. Kestelman.

4.175 *AMM 85*, 1978, 749–750, R. P. Boas, H. Pollard, D. Widder.

4.176 **Hint:** denote $\Delta f(x) = f(x + h) - f(x)$. Then $\Delta^n f(x) = h^n f^{(n)}(x + t)$ for some $t \in (0, nh)$.
AMM 89, 1982, 452, Putnam Competition 1958, PM4, H. Flanders.

4.177 **Hint:** take the logarithms. Observe that if all derivatives of a function g at 0 are positive, then e^g has the same property.
[MS] 179, 4.

4.179 *AMM 76*, 1969, 304–305, E2069, R. P. Sheets; K. Miller.

4.180 **Answer:** $A = x^2$, $B = xd/dx - 1$.
[MS] 121, 1.

4.181 **Hint:** approximate f by convex functions.
AMM 70, 1963, 677, 5032, A. Wilansky; J. H. Michael.

4.183 **Hint:** use the Laplace transform.

4.186 **Hint:** estimate $f(c) - f(t)$ and $f(-c) - f(t)$ using the Taylor formula.
[Di] 8.14.2.

4.187 [Di] 8.12.3.

4.188 For example *AMM 87*, 1980, 805–809, S. Roman.

4.189 *NAW 3*, 1985, 281–283, A. Abian.

4.190 **Remark:** for $f(x) = x^n + x^{n+p}$ the expression $(c - a)/(b - a)$ is constant.
AMM 89, 1982, 311–312, A. G. Azpeitia.

4.192 Hint: if $F(x) = \Pi_{k=1}^{n}(x - r_k)$, then $P(d/dx)f = \Pi_{k=1}^{n-1}(d/dx - r_k)$ $\times (d/dx - r_n)f = \Pi_{k=1}^{n-1}(d/dx - r_k)(f' - r_n f)$. Then use the result from 4.142.
AMM 64, 1957, 23, Putnam Competition 1956, I-4.

4.193 *EM 39*, 1984, 100–102, 898, M. S. Klamkin, A. Meir; P. Bundschuh.

4.194 *AMM 69*, 1962, 1015–1016, 4989, Y. Matsuoka, R. L. Van de Wetering.

4.195 *EM 40*, 1985, 154, 933, H.-J. Seiffert.

4.197 Hint: show that if $f^{(n+1)}$ has no zeros, then $f^{(n)}(x)$ tends to 0 as x tends to $\pm\infty$.
Remark: the conclusion is optimal, for example for $f(x) = \int_{-\infty}^{x} \int_{-\infty}^{t_n} \cdots \int_{-\infty}^{t_2} \exp(-t_1^2)\,dt_1 \cdots dt_n$ $f^{(k)}(x) \neq 0$ for $0 \leqslant k \leqslant n$.
AMM 87, 1980, 222–223, E2755, M. Slater; R. L. Cooke.

4.198 For a far-reaching multidimensional generalization (the Grobman-Hartman theorem) see for example J. Palis, W. de Melo, *Geometric Theory of Dynamical Systems*, Springer, New York, 1982, Ch. 2, Section 4.

4.199 Hint: if $n(x)$ are uniformly bounded then the proof is easier.
Answer: for $f(x) = \sin(x/2)$ $\lim_{n\to\infty} f^{(n)}(x) = 0$.
Rev. Mat. Hisp.-Amer. 4, 14, 1954, 26–43, E. Corominas, F. Sunyer Balaguer; [K12] 5.11; *J. Math. Anal. Appl. 140*, 1989, 301–309, A. Boghossian, P. D. Johnson, Jr.

4.201 Answer: the functions sin and cos are not similar. These quadratic functions are similar if $4b = a^2 - 2a$.

4.202 [K-G] 478.

4.203 Hint: for every open set $G \subseteq \mathbb{R}$ there exists a smooth function which is strictly positive exactly on G. If $G = \bigcup_n (a_n, b_n)$ with mutually disjoint intervals (a_n, b_n) then define $f_n(x) = \exp(-(x - a_n)^{-2}(b_n - x)^{-2})$ for $a_n < x < b_n$ and 0 elsewhere. Let $M_n = \sup\{|f_n^{(k)}(x)| : 0 \leqslant k \leqslant n,\ x \in \mathbb{R}\}$, $f = \Sigma_n f_n/n^2 M_n$.
CMB 12, 1969, 25–30, L. E. May.

4.204 Answer: not always. See the preceding problem.

4.205 Hint: use function defined in the hint to 4.203 with suitably chosen (a_n), (b_n).
AMM 91, 1984, 618–624, F. S. Cater. See 5.47.

4.206 Answer: yes.
Remark: this theorem can be generalized to the case of functions defined on $\mathbb{R}^n$ of class C^m, $m \in \mathbb{N} \cup \{\infty\}$, with support in fixed open set.

AMM 84, 1977, 303–304, 6042, F. T. Laseau; G. M. Leibowitz, C. H. Rasmussen, S. J. Sidney, J. Boman.

4.207   *AMM 74*, 1967, 867–868, E1875, S. Zaidman; S. Philipp.

4.209   **Hint:** study the differentiability of the rest in the Taylor formula.
Answer: for an odd function f $f(x) = xg(x^2)$. If f is in C^{2k+1} then g belongs to C^k; if f is analytic then g is also analytic.
Duke Math. J. 10, 1943, H. Whitney; [B-L] 4.18.2.

4.210   [Di] 8.14.4; other proofs: [N1], [R2], [B-L].

4.211   **Answer:** yes.
AMM 94, 1987, 695, 6374, Lee Witt; see *Arch. Math. 39*, 1982, 269–277, H. Joris; *J. Analyse Math. 45*, 1985, 46–68, J. Duncan, S. G. Krantz, H. R. Parks.

4.212   *AMM 90*, 1983, 557–560, M. J. Hoffman, R. Katz; [B] 183, *Math Z. 62*, 1955, 354–367 (Hilfsatz 6), H. Salzmann, K. Zeller; D. Widder, *The Laplace Transform*, Princeton 1946, 146.

4.213   **Hint:** write the rest in the Taylor formula as $R_n(x) = \int_0^x \int_0^{t_n} \cdots \int_0^{t_2} f^{(n)}(t_1) \, dt_1 \cdots dt_n$, $n \geqslant 2$. Observe that $R_n(x)/x^n$ increases, $0 \leqslant R_n(r) \leqslant f(r)$ and $0 \leqslant R_n(x) \leqslant (x/r)^n f(r)$.
AMM 90, 1983, 130–131, R. Redheffer; [G] 464. See also the last reference to 4.199.

4.214   **Hint:** if the derivatives of f are large, then they would have many zeros in I, see 4.187.
[Di] 9.9.7.

4.215   **Hint:** use the Bernstein theorem, see 4.213.
Answer: no.
AMM 91, 1984, 60, 372, 6392, C. W. Henson, B. Reznick, L. A. Rubel; P. R. Chernoff; *AMM 92*, 1985, 364.

4.216   For example R. R. Phelps, *Lectures on Choquet's Theorem*, D. Van Nostrand, Princeton, 1966, Ch. 2; [Fe] II, XIII, Section 4.

4.217   **Answer:** for example $f(t) = (1 + t^2)|\sin \pi t|(\log |\sin \pi t|)^2$; the important properties are: $f(n) = 0$ for $n \in \mathbb{Z}$, $f(t) > 0$ elsewhere and $\int_{\mathbb{R}} 1/f < \infty$.
AMM 89, 1982, 241–244, W. Rudin.

4.219   *AMM 93*, 1986, 743, 6530, W. Walter.

4.220   **Hint:** a function f satisfies the Lipschitz condition with the constant 1 if and only if $x + f(x)$ and $x - f(x)$ are nondecreasing.
AMM 95, 1988, 667–668, 6529, O. Hájek; M. J. Pelling.

Answers to Chapter 5

Functional Equations and Functions of Several Variables

5.1 **Hint:** observe that if $x_0 = x$ and $x_{n+1} = (x_n - 1)/2$ then $\lim_{n \to \infty} x_n = -1$.
Answer: Constant functions are the only solution.
MS, 4, 1978.

5.2 **Answer:** f is a constant function.

5.3 *MS, 5,* 1977.

5.4 **Answer:** affine functions.

5.5 **Answer:** no.
AMM 75, 1968, 915, 5530, S. S. Mitra, J. R. Porter, W. J. Mourant.

5.6 **Answer:** $f(x) = (x^3 - x^2 - 1)/(2x(x - 1))$.
AMM 80, 1973, 173, Putnam Competition 1972, B-2.

5.7 **Answer:** the only solution is

$$f(x) = \frac{1}{2}(x - 1/(1 - x) + 1 - 1/x)$$

5.8 **Answer:** $f \equiv 0$ and $f(x) = -x^n(x - 1)^n$.

5.9   Answer: for $f(x) = qx + r$, where q is rational.

5.10   Answer: (i) the identity is the only such homeomorphism.
Remark: the continuity and bijectivity assumptions are superfluous.
AMM 89, 1982, 702–703, E2893, Ko-Wei Lih; F. Richman, I. M. Isaacs, D. Laugwitz.

5.11   *AMM 93*, 1986, 818, E3053, I. C. Bivens; V. Hernández; see the preceding problem.

5.12   Answer: only constant functions.
AMM 84, 1977, 388, E2575, D. Shelupsky; A. Nijenhuis.

5.13   Answer: $f(\sqrt{q}) = \sqrt{q}$.
MS, 2, 1978.

5.14   Hint: show that if $f(1) \neq 0$ then $f(1) = 1$, hence $f(x) = x$ for rational x.
Observe that f is strongly increasing.
Answer: in the complex case either $f \equiv 0$ or $f(z) = z$ or $f(z) = \bar{z}$.
[Di] 4.3.1, 4.4.1.

5.15   Hint: (i) for $c \in (a, b)$ consider x, y such that $x < c < y$, $(y - c)/(c - x) \in \mathbb{Q}$.
(ii) show that f satisfies the equation

$$yf(x) = \int_0^{x+y} f - \int_0^x f - \int_0^y f = xf(y)$$

(iii) observe that if f would not be linear then for some real a the sets $\{x > 0 : f(x)/x \geqslant a\}$, $\{x > 0 : f(x)/x < a\}$ would be of strictly positive measure and they (completed by 0) would be the additive semigroups. Apply the Steinhaus theorem.

For example [Di] 4.1.1; *AMM 80*, 1973, 1041, H. Shapiro; *AMM 85*, 1978, 663–664, G. Letac; see 6.14.

5.16   Answer: if for example $f(x)f(1/x) \geqslant a > 0$, then f is continuous. However there exists discontinuous $f : \mathbb{R} \to \mathbb{C}$ satisfying these properties: $f(x) = cx + id(x)$, where $c \in \mathbb{R}$ and $d(x)$ is a discontinuous derivation in $\mathbb{R}$. Then $f(x)f(1/x) = c^2 + d^2(x)/x^2 > 0$.
Aequationes Math. 19, 1979, 300, P. 178, F. Rothberger.

5.17   Hint: integrate f over the interval $[-t, t]$ and write a differential equation for this integral.
Answer: $f \equiv 0$, $f \equiv 1$, $f(x) = \cos kx$, $f(x) = \cosh kx$, where k is an arbitrary real number.

5.18   Hint: change the variables to get the Cauchy functional equation.
[Hi1] 14.2.16.

5.20   **Answer:** $f(x) = kx^2/2 + cx$, where $c = f(1) - k/2$.

5.21   **Hint:** in the first part show that $f(x + a) = g(a)f(x) + h(a)$, where $g(a) \neq 0$ for every a. Then prove that f and g are continuously differentiable and $f'(x) = cf(x) + d$.

The second part of the problem may be reduced to the first one by a change of variables.
[Hi2] 15.2.1–15.2.4.

5.22   **Hint:** show that f satisfies the equation $f(xy) = f(x)f(y)$.
For example *AMM 75*, 1968, 646, T. G. Newman.

5.23   **Answer:** $F(x, y) = g(x) - g(y) + a(x + y) + b(x^2 + 4xy + y^2) + c$,   where g is an arbitrary continuous function and a, b, and c are real constants.
AMM 84, 1977, 824–825, E2607, J. A. Baker.

5.24   *AMM 84*, 1977, 551–553, S. Gudder, D. Strawther.

5.25   *AMM 86*, 1979, 394, E2677, E. Just; G. W. Peck.

5.26   **Hint:** consider the functions $(f(x) \pm f(1/x))/2$.
AMM 85, 1978, 208–210, 6106, D. S. Mitrinović, P. M. Vasić; C. Henley.

5.27   **Answer:** yes.
AMM 85, 1978, 200–201, E2626, R. Johnsonbaugh; P. Erdös.

5.29   **Answer:** (i) $f(x) = c \log |x|$, (ii) $f(x) = c \log x$ with an arbitrary constant c.
[CHM] 5/1957, R.13.

5.31   **Answer:** $f(z) = Az$, $f(z) = A \sinh bz$, $f(z) = A \sin bz$, $A \in \mathbb{C}$, $b \in \mathbb{R}$.
For example, *AMM 64*, 1957, 83–85, R. M. Robinson.

5.32   [Hi1] 6.4.2.

5.33   [Hi1] 6.4.10.

5.34   **Hint:** take as the argument of the primitive of f $x = \exp(\exp y)$.
AMM 75, 1968, 412–413, E1947, U. Dudley; C. J. Henrick.

5.35   *AMM 75*, 1968, 734, Putnam Competition 1967, A-4.

5.37   For example *AMM 70*, 1963, 893, E1549, C. R. MacCluer; D. A. Moran.

5.38   **Hint:** the solution may be obtained using the Picard iteration method with $F_0 = 0$.
AMM 68, 1961, 73, 4885, M. S. Klamkin, D. J. Newman; P. G. Rooney.

5.39   **Answer:** the only solution is $f(x) = a^{-1/a} x^a$ where $a = (1 + \sqrt{5})/2$.

5.40   **Answer:** $y(x) = \dfrac{1}{(n-1)!} \displaystyle\int_1^x (x - t)^{n-1} g(t) t^{-n}\, dt.$

AMM 71, 1964, 636, Putnam Competition 1963, I-3.

5.41   *AMM 94*, 1987, 683–684, E3055, M. F. Kruelle; O. P. Lossers.

5.42   *AMM 84*, 1977, 295–296, E2568, S. Bell; S. R. Glidewell, J. F. Grimland.

5.43   **Answer:** $y(x) = e^x = \Sigma_{k=0}^{\infty} x^k/k!$ and $y_n(x) = \Sigma_{k=0}^{n} x^k/k!$.
AMM 74, 1967, 436, J. P. Hoy; *AMM 84*, 1977, 716–720, D. Zeitlin.

5.44   **Answer:** the only solution is $f(x) = (\log(x^2 + e))^{1/2}$.
[MS] 159, 54.

5.45   **Answer:** $y(t) = (2e^t + 1 - e)/(3 - e)$.
[MS] 127, 4.

5.46   **Answer:** for example $f(x) = \Sigma_{n=-\infty}^{\infty} (-1)^n c^{-n(n+1)/2} \exp(-c^n x)$ satisfies these conditions.

5.47   **Hint:** solve an integral equation satisfied by f. Observe that f is smooth but nowhere analytic.
A modification of *AMM 92*, 1985, 514–515, 6451, M. C. Mallows; T. Bang.

5.48   **Hint:** f is the tempered distribution so $\hat{f}$ is. Another method: show that f is analytic.
Answer: there are functions of exponential growth e^{zx}, with some complex z, satisfying this mean value property. See 7.88.
   For further generalizations see *AMM 81*, 1974, 115–137, L. Zalcman, and *Arch. Rational Mech. Anal.* 47, 1972, 237–254.

5.49   **Answer:** constant functions, bijections and even functions which are bounded and strictly monotone on $[0, \infty)$.
AMM 84, 1977, 571, E2583, C. L. Mallows, P. L. Montgomery.

5.50   **Hint:** consider $f(x, y) - f(x - 1, y)$ or use Lagrange interpolation polynomials.
Answer: this is not true for functions defined on $\mathbb{Q} \times \mathbb{Q}$.
For example *AMM 91*, 1984, 142, E2940; *AMM 68*, 1961, 42, F. W. Carroll; *CMB 19*, 1976, 252–253, P. 238, E. Grassmann.

5.51   *AMM 91*, 1984, 114–116, P. G. Laird, R. McCann.

5.52   Sketch of the proof:
(i) For every system of vectors $X = (x_k)$, $k = 1, \ldots, m$, such that $\Sigma_{k=1}^{m} x_k = 0$, $|x_k| \leqslant 1$, there is a permutation $x_{j_1}, \ldots, x_{j_m}$ satisfying $|\Sigma_{k=1}^{r} x_{j_k}| \leqslant C$ for

$r = 1, \ldots, m$, where C does not depend on X (a proof by induction on $\dim(\operatorname{lin} X)$).

(ii) F contains the cluster points of the partial sums of $\Sigma\, x_k$.

(iii) F is a linear subspace different from $\{0\}$.

(iv) If the domain of absolute convergence of the series $\Sigma\, x_k$ is $\{0\}$, then $F = \mathbb{R}^n$.

[G-L] VI, 152–157; *AMM 94*, 1987, 342–351, P. Rosenthal. For related results in the infinite dimensional case see V. M. Kadec, *Funkc. anal. i ego pril.* 20, 4, 1986, 74–75, and *Izv. VUZ Matematika*, 12, 1986, 32–36.

5.53 Answer: no in the first case, for example $f(x, y) = (x - c)(y - c)$, $0 \leqslant c \leqslant 1$. Yes in the second case.
AMM 89, 1982, 277, E2854, J. Chow; D. M. Bloom.

5.54 Hint: the functions

$$i(t) = \begin{cases} \min\{f(x) : g(x) \geqslant t\} & \text{if } t \leqslant \max\{g(x) : x \in P\} \\ \max\{f(x) : x \in P\} & \text{if } t > \max\{g(x) : x \in P\} \end{cases}$$

and $j(t) = t^{-1} \int_0^t i(u)\, du$ may be useful.
AMM 66, 1959, 148, 4756, J. L. Massera; J. Rainwater.

5.55 Answer: $a, b \geqslant 0$, $a + b = 1$.

5.56 Hint: compute $Q(D)Q(x)$ for polynomials Q. Observe that if $P(D)(A(x)P(x)) = 0$ then also $A(D)P(D)(A(x)P(x)) = 0$.
AMM 69, 1962, 674, 4965, H. S. Shapiro; R. M. Thrall, J. G. Wendell.

5.57 Hint: consider a continuous function $g : \mathbb{R} \to \mathbb{R}$ such that $f(e^{2\pi i t}) = e^{2\pi i g(t)}$. Then the differences $g(2t) - 2g(t)$ and $g(t+1) - g(t)$ are integers. Hence $g(t) = nt + m$ for some integers n, m.
Answer: $f(z) = z^n$, $n = 0, 1, 2, \ldots$.
AMM 87, 1980, 581, E2783, W. Knight, B. Lund; I. M. Isaacs.

5.58 *AMM 87*, 1980, 679, 6250, H. Shapiro; C. L. Belna.

5.59 Answer: the fixed point: $(0, 0)$, the periodic points (of period 2): $(2, -4)$, $(-2, 4)$.
[Hi2] 5.4.3.

5.60 Hint: suppose that $a = 0$ and consider the map $x \to x/|x|^2$, $x \neq 0$.
[Di] 6.2.4.

5.61 Answer: the minimal value is obtained for $r = \sqrt{2}$, $s = 2$, $t = \sqrt{8}$.
AMM 89, 1982, 681, Putnam Competition 1981, B-2.

5.62 Hint: the problem may be reduced to a two-dimensional one.
Answer: there are six minima and eight maxima.
[C] 1.31.

5.63   **Hint:** consider the polynomial $(t - x_1) \cdots (t - x_n)$.

5.64   **Hint:** the problem reduces to the study of the function $f(x, y) = \sin^p x + \sin^p y + \sin^p(x + y)$ where $x, y, x + y \in [0, \pi]$.
Answer: for $p \neq 4$ the extremum is attained for the equilateral triangle; for $p > 4$, it is a local minimum, for $0 < p < 4$ it is a local maximum. For $p = 4$, $x = y = \pi/3$ is not a local extremum.
[Hi2] 15.4.2.

5.65   **Hint:** the de l'Hôpital rule may be useful.
Answer: this limit is infinite.

5.66   **Hint:** consider the new variables $s = u - v, t = u + v$ and apply the de l'Hôpital rule.
Answer: the limit is equal to 2/9.

5.68   [MS] 147, 36.

5.69   **Hint:** apply the Hölder inequality.
AMM 90, 1983, 411–412, 6372, D. R. Brillinger; Pei Yuan Wu.

5.70   **Hint:** $I = \int \cdots \int f(0 - x_n) f(x_n - x_{n-1}) \cdots f(x_2 - x_1) f(x_1) \, dx_1 \cdots dx_n$
$= (2\pi)^n \int e^{ixt} (\hat{f}(t))^{n+1} \, dt|_{x=0}$.
AMM 87, 1980, 53–54, H. Dym; *AMM 82*, 1975, 672–673, P. Ungar; *AMM 94*, 1987, 83–86, 6493, M. L. Glasser; W. A. Newcomb, A. A. Jagers.

5.71   **Hint:** the problem has a nice probabilistic interpretation.
Answer: 1/3.
[MS] 141, 28.

5.72   **Hint:** diagonalize the matrix A and after a change of variables compute the integral of $\exp(-\Sigma_{k=1}^{N} z_k^2 - i(z, Cz))$ diagonalizing the matrix C.
For example [Be] 6.9.1.

5.73   **Hint:** consider first $f(x, y) = x^k y^n$.
Answer: the limit is equal to $f(1/2, 1/2)$.
[MS] 166, 4.

5.74   **Answer:** the limit is equal to

$$\frac{1}{2} \int_0^1 (f(1, y) - f(0, y)) \, dy + \frac{1}{2} \int_0^1 (f(x, 1) - f(x, 0)) \, dx$$

[MS] 166, 7.

5.75   **Answer:** the limit is equal to

$$\frac{1}{2\pi} \int_a^b \int_0^{2\pi} f(x, \sin y) \, dx dy$$

Remark: this generalizes the Fejér theorem. See 4.126, 8.19.
[MS] 152, 44; 180.

5.76 [MS] 187, 5.

5.79 **Answer:** no. A counterexample:
$$f(x, y) = \begin{cases} \exp(-x^2/y^2 - y^2/x^2) & \text{for} \quad xy \neq 0 \\ 0 & \text{for} \quad xy = 0 \end{cases}$$

5.80 **Answer:** the equality $(d^2y/dx^2)(d^2x/dy^2) = 1$ implies $(d^2y/dx^2)^2 + (dy/dx)^3 = 0$ hence $(x + a)(y + b) = 4$ for some a, b.
AMM 71, 1964, 556, E1612, C. S. Ogilvy; E. S. Eby.

5.81 *AMM 75*, 1968, 787, E1986, H. S. Shapiro.

5.82 **Answer:** no, $f'(\mathbb{R})$ may be disconnected: for example $f(t) = (t^2 \sin(1/t), t^2 \cos(1/t))$ for $t > 0$, $f(t) = (0, 0)$ for $t \leq 0$.
[Di] 8.5.4.

5.83 **Answer:** no, since $\partial f/\partial x$, $\partial f/\partial y$ are of the first Baire class they are continuous on some dense G_δ sets. Df exists on their intersection.

5.85 **Hint:** use a polygonal approximation.
[K-G] 318.

5.86 [N1] 2.5.10, 2.5.11.

5.87 [B-L] 3.4.3.

5.88 **Answer:** for $n = 1$ only.
AMM 93, 1986, 307, 6483, J. Arias de Reyna; O. Jørsboe.

5.89 **Answer:** $P_{a,b}$ has local minima at $(a, 0)$ and $(b, 0)$ and a saddle point at $(0, 0)$; $P_{a,b}(a, 0) \neq P_{a,b}(b, 0)$ if $a \neq -b$. $P_{a,b}^{-1}((-\infty, c])$ is connected if $c \geq 0$; for $c > 0$ this set is homeomorphic to the k-dimensional ball.
[Hr] 9.2.2.

5.90 **Hint:** consider the perturbed function $g(x) = f(x) + \varepsilon c^{-1}|x - x_\varepsilon|$.
AMM 90, 1983, 206–207, J.-B. Hiriart-Urruty.

5.91 **Hint:** (i) represent the complement of this closed set as a countable union of balls and construct smooth positive functions supported on these balls. Then sum up these functions multiplied by sufficiently small positive numbers to control the convergence of the derivatives.
[B-L] 3.3, 3.4.2.

5.92 **Answer:** let $g(x) = 0$ on the Cantor set and $g(x) = (x - a)(b - x)$ on the intervals (a, b) deleted in the construction of the Cantor set. Put $f(x) = \int_0^x g(t)\, dt$.
AMM 71, 1964, 693, 5114, W. E. Johnson, C. M. Petty; W. C. Waterhouse.

5.93 *AMM 84*, 1977, 722–723, W. Rudin.

5.94 *AMM 84*, 1977, 162–167, G. R. Sell.

5.95 Answer: no. For example $f(x,y) = xy(-\log(x^2+y^2))^{1/2}$, $f(0,0) = 0$. Another example is given in *Dokl. AN SSSR 123*, 4, 1958, 606–609, B. S. Mitiagin. For a general setting see *Acta Math. 31*, 1908, 127–332, and *J. Math. Pures Appl. 5*, 1909, 127–233, H. Petrini.

5.96 Hint: computing formally $f(x, y - x)$ does not depend on x.
Answer: yes.
Remark: the problem in the general setting is difficult.
AMM 85, 1978, 829, 5871, P. R. Chernoff; *AMM 80*, 1973, 1150–1151, K. F. Anderson; *AMM 82*, 1975, 530–531, H. Royden; *AMM 83*, 1976, 573, M. J. Pelling.

5.97 Hint: consider $\min_{x,y} ((u - x - f(y))^2 + (v - y - f(x))^2)$ for arbitrary u and v.
[C] 1.30.

5.99 Answer: no.
AMM 86, 1979, 230–231, 6167, C. R. Williams, J. C. Warndorf; W. Emerson, J. Komló, R. Pollack.

5.100 [Hr] 2.1.10.

5.101 *AMM 91*, 1984, 145–146, 6397, E. Feldman, M. Vulis; B. Schaaf.

5.102 Hint: consider $g(y) = (y - F(y))/|F(y) - y|$ and apply the Brouwer theorem.

5.103 Hint: reduce the problem to this one with differentiable mappings. Consider the auxiliary transformation $F_t(x) = (1 - t)x + tF(x)$ which is bijective on the unit ball for small t. Show that the Jacobi determinant (which is a polynomial in t) is constant.
AMM 85, 1978, 521–524, J. Milnor; *AMM 86*, 1979, 572–574, M. Eisenberg, R. Guy; *AMM 87*, 1980, 525–527, C. A. Rogers; *AMM 88*, 1981, 264–268, Y. Kannai.

5.104 Hint: denoting $F = (f, g, h)$ and $G = x|x|^2$, observe that $G^{-1} \circ F$ is bounded and has a fixed point.
NAW 13, 1965, 131, 12, N. G. de Bruijn; A. E. Livingston.

5.105 Hint: show that $x \mapsto x \circ x$ would be a homeomorphism which is absurd: $(-x) \circ (-x) = x \circ x$. Remember that for $n \geqslant 3$, $\mathbb{R}^n \backslash \{0\}$ is simply connected.
Remark: the Hadamard (global) implicit function theorem may be recalled here: If $F : M_1 \to M_2$ is a C^1 mapping between connected n-dimensional manifolds with nonvanishing Jacobi determinant, M_2 is simply connected

and $F^{-1}(C)$ is compact whenever C is compact, then F is the homeomorphism.
AMM 84, 1977, 28–29, W. B. Gordon.

5.106 Hint: prove by induction that the hypotheses of the inverse function theorem are satisfied. Apply the Viète formulas.
[S1] 1.59.

5.107 Hint: the Sard theorem and the result of 5.91 may be useful.
[N1] 1.4.12, 1.4.14.

5.108 [N1] 1.4.16.

5.109 Hint: express the Fourier transform of f at d using the values of the integrals of f along the lines perpendicular to $\vec{d}$.
For example *AMM 64*, 1957, 750–751, 4721, D. J. Newman; *AMM 85*, 1978, 601–602, 6120, J. Fishburn; K. J. Falconer.

5.110 Hint: it suffices to compute $f(0) = \int \hat{f}(a)\,da$. Similarly as in the preceding problem $\hat{f}$ may be expressed in terms of $F(L)$ for $L = \{x : (a, x) = b\}$, $b \in \mathbb{R}$.
Remark: the result in odd-dimensional case is similar as for $n = 3$.
[K-G] 689.

5.111 [K-G] 690.
Supplementary references: S. Helgason, *The Radon Transform*, Birkhäuser, 1980; *AMM 89*, 1982, 377–384, 420–423, R. S. Strichartz; *AMM 87*, 1980, 161–175, L. Zalcman; *AMM 85*, 1978, 420–439, L. A. Shepp, J. B. Kruskal; [S2] Section 11.

5.112 Hint: recall the properties of zeros of Bessel functions.
[K-G] 702.

5.113 [C] 2.3.

5.114 Answer: $f(x_1, \ldots, x_n) = \int_0^1 h(tx_1, \ldots, tx_n)t^{p-1}\,dt$.
[C] 3.10.

5.115 Answer: $u(x, y) = (c_1 + c_2x)e^{ax}e^{ay}$, $u(x, y) = (k_1 + k_2x)e^{bx}e^{by}$.
[CHM] 7/1961, R.14.

5.116 Answer: for affine functions only.

5.120 Hint: if $u(z_0) = 0$ for some $z_0 = x_0 + iy_0$, $|z_0| = R$, then $u(z) \geqslant (R - |z|)/(R + |z|)$ for $|z| \leqslant R$ from the Harnack inequality. For the real functions f and g defined by $f(t) = u(tz_0)$, $g(t) = (1 - t)/(1 + t)$, $f(t) \geqslant g(t)$ holds for $0 \leqslant t \leqslant 1$, $f(1) = g(1) = 0$ and $f'(1) \leqslant g'(1)$; hence $xu_x + yu_y \leqslant -1/2$ at z_0. In other words, for the harmonic function u the conditions $u(0) = 1$, $u(z) = 0$, imply $xu_x + yu_y > -1/2$ so $u > 0$.

A generalization: for every $u \in C^1(D)$ and g such that $g(s) \leqslant 0$ for $s \leqslant 0$, if $g(u) + xu_x + yu_y > 0$, then $u > 0$.
AMM 86, 1979, 710–711, 6198, S. S. Miller, W. Walter.

5.121 **Answer:** not always. The example may be constructed observing that the strictly positive harmonic function F in D satisfies the Harnack inequality: $F(z)/F(w) \leqslant 12$ if $|w| \leqslant 1/2$, $|z| \leqslant 1/2$.
AMM 91, 1984, 61, 6393/I, G. A. Edgar; I. Netuka, J. Vesely.

5.122 **Hint:** show that the weak maximum principle applies to the function with restricted mean value property. Then use the assumption on the solvability of the Dirichlet problem.
AMM 87, 1980, 819–820, R. B. Burckel.
The original paper: *Trans. AMS 36*, 1934, 227–242, O. D. Kellogg. See also *Arch. Rational Mech. Anal. 47*, 1972, 237–254, L. Zalcman and the references to 5.111.

5.123 **Answer:** yes, for example $f(x_1, \ldots, x_n) = \operatorname{sgn} x_1$ satisfies all the conditions but the continuity.
Remark: A similar result holds for the spherical means.
T. Radó, *Subharmonic Functions, Ergebnisse der Mathematik*, Springer, 1937, 5, 14.
More generally: M. Brelot, *Théorie classique du potentiel*, Centre de documentation universitaire, Paris 1969, Ch. 2; *AMM 86*, 1979, 229–230, 6165, A. G. O'Farrell; R. L. Cooke.

5.125 **Answer:** $\{(u, v): u^n + v^n = 1, 1/n + 1/m = 1\}$.

5.126 **Answer:** this is the parabola $x = -y^2/4$.
[Hi2] 13.2.6.

5.127 **Answer:** the trajectory is a cycloid.
AMM 64, 1957, 23, Putnam Competition 1956, I-3.

5.129 *NAW 26*, 1978, 235, 468, O. Bottema; H. Kappus.

5.130 *AMM 73*, 1966, 542, E1770, G. P. Graham, T. A. Porsching; W. Baron.

5.132 *AMM 76*, 1969, 34–35, P. R. Chernoff.

5.134 *AMM 72*, 1965, 678, 5211, R. Redheffer, W. C. Waterhouse.

5.135 **Answer:** $y(x) = k/x$ with $k > 0$.

5.136 **Answer:** for the ellipse this is the curve
$$((a^2 - b^2)a^{-1} \cos^3 t, (b^2 - a^2)b^{-1} \sin^3 t)$$
[Th] 16.1.

5.137 Hint: Look for the parametric equation of this curve. Begin with a sketch of the curve (it has four cusps).
Remark: It is interesting to consider other values of a and z.
[S-P] 560.

5.138 Hint: $f(P) - f(Q) = (x_P - x_Q)f_x + (y_P - y_Q)f_y + (z_P - z_Q)f_z$.
AMM 64, 1957, 195–196, R. Redheffer.

5.139 Hint: consider a simple fraction representation of $F(x) = ((x - a)^2 + b^2)/f(x)$ where $P = (a, b)$, multiply by x, and take the limit for $x \to \infty$.

5.140 Answer: $P(1,0) = 1$ and $P(-1,0) = -1$; hence the boundary of $\{x^2 + y^2 \leqslant 1\} \cap \{P(x, y) > 0\}$ contains a curve C connecting two points A and B in the upper and lower semicircle. $Q(A) > 0, Q(B) < 0$ so there exists a point on C where Q vanishes.
AMM 69, 1962, 574, 4962, D. J. Newman; R. Breusch.

5.141 AMM 69, 1962, 240–241, 4948, L. Glasser; D. J. Newman.

5.142 Hint: $\pi/2e = \lim_{N \to \infty} \Pi_{n=1}^{2N} (1 + 2/n)^{(-1)^{n+1}n}$.
AMM 68, 1961, 39–41, Z. A. Melzak.

Answers to Chapter 6

Real Analysis, Measure and Integration

6.1 Answer: for (the most difficult case of) the closed interval $[-2, 2]$, $n = 1, 2, \ldots$, the partition is given by $\bigcup_{j=0}^{\infty} I_{k,2j} \cup \bigcup_{j=0}^{\infty} I_{k+1,2j+1}$, where $I_{0,0} = [-2, 2^{-n+1}]$, $I_{0,j} = (1 + 2^{-j}, \ 1 + 2^{-j+1}]$, $I_{k,0} = [2^{-k} + 2^{-k-1}, 2^{-k+1}]$, $I_{k,j} = [2^{-k} + 2^{-k-j-1}, \ 2^{-k} + 2^{-k-j})$, $k = 1, \ldots, n-1$, $I_{n,j} = I_{0,j}$, $n = 2, 3, \ldots, j = 0, 1, \ldots$. For $n = \omega$: $I_{0,2k-1} \cup \bigcup_{j=0}^{\infty} I_{k,2j} \cup \bigcup_{j=0}^{\infty} I_{k+1,2j+1}$, where $I_{0,-1} = \varnothing$, $I_{0,0} = [-2, 0]$.
AMM 87, 1980, 406–407, E2768, L. F. Meyers.

6.2 Answer: no. Given an arbitrary dense set in the plane one can select a countable dense subset which is not dense in any segment.
Remark: W. Sierpiński has proved in *Fund. Math. 1*, 1920, 112–115, that there exists a nonmeasurable subset of the plane which meets each closed set of positive measure but which intersects each line in at most two points.
AMM 67, 1960, 189, 4843, L. Flatto; L. Moser.

6.4 Hint: denote $a_n(x) = (\chi_{E_1}(x) + \cdots + \chi_{E_n}(x))/n$, $f_n(x) = \sup_{m \geqslant n} a_m(x)$, $f(x) = \limsup_{n \to \infty} a_n(x) = \lim_{n \to \infty} f_n(x)$. Show that there is an x such that $f(x) \geqslant c$. If no such x would exist then $c \leqslant \limsup_{n \to \infty} \int_0^1 a_n(x)\, dx \leqslant \lim_{n \to \infty} \int_0^1 f_n(x)\, dx < c$.
AMM 71, 1964, 105, 5074, P. Erdös; R. Scoville.

6.5 **Hint:** define first T on E by $T(x) = \int_0^x \chi_E(t)\, dt$.
AMM 76, 1969, 97, 5557, P. R. Chernoff.

6.6 *AMM 68*, 1961, 381–382, E1434, A. Beck; 668, R. G. Kayel.

6.7 **Answer:** the union of M may be nonmeasurable and the length of M may exceed c.
Remark: an interesting result on nonmeasurable unions of sets of measure zero is in *Bull. Pol. Acad. Sci. Math. 27*, 1979, 447–448, J. Brzuchowski, J. Cichoń, E. Grzegorek, C. Ryll-Nardzewski.

6.8 [G] 244.

6.10 [G] 274, 457.

6.11 **Hint:** the Baire category theorem applied to the Fréchet-Nikodym space of the measure m may be useful.
[Te] 6.90; see *AMM 91*, 1984, 564–566, F. S. Cater.

6.12 [H-S] 18.30, 18.31.

6.13 **Corollary:** there is no measurable set $A \subseteq [0,1]$ such that $m(A \cap I) = m(A^c \cap I) = m(I)/2$ for every open interval $I \subseteq [0,1]$.
See other versions of this result: [G] 230, 245, 246; [K-G] 246; [O] 441; *AMM 71*, 1964, 328–329, 5083, J. Barlaz; W. J. Bruecks; *AMM 80*, 1973, 209–210, 5821, E. Langford; J. W. Grossman; *AMM 89*, 1982, 114–116, 125, A. Simoson; *AMM 91*, 1984, 190–193, A. Simoson; *AMM 90*, 1983, 41–42, W. Rudin.
See also related results: *AMM 91*, 1984, 19–22, H. Diamond, G. Gellès; *AMM 90*, 1983, 403, 6435, Th. Q. Sibley.

6.14 *CMB 21*, 1978, 497–498, D. Borwein, S. Z. Ditor; *Proc. AMS 105*, 1989, 889–893, H. Miller.

6.15 **Hint:** consider the dyadic intervals.
Another possibility: apply the Radon-Nikodym theorem or the Liapunov theorem on the convexity of the image of the vector measure (m, μ).
AMM 87, 1980, 411, 6242, J. Mycielski; E. Bolker, F. S. Cater, R. C. Shiflett.

6.16 [G] 249.

6.17 [G] 240.

6.18 [R3] 9.9.

6.19 [G] 205.

6.20 **Hint:** consider for the Borel set $X \subseteq \mathbb{R}$ $Y = \{(x_1, x_2) : x_1 - x_2 \in X\}$.
Apply the Fubini theorem for $\mu \times m$.
[K-G] 243.

6.21 Answer: no, although this is true if the measure of $[0, 1]$ is finite. Then this measure is a multiple of the Lebesgue measure.
Hint: construct a Borel set B such that $-B \not\subset \bigcup \{B + a_n : n \in \mathbb{N}\}$, where (a_n) is a real sequence.
AMM 90, 1983, 62–63, 6348, J. Henle, S. Wagon; *Proc. Cambridge Phil. Soc.* 62, 1966, 693–698, K. E. Hirst.

6.22 Answer: consider for example the function f which assigns to the rectangle P the length of the intersection of P with a fixed diagonal of the square divided by $\sqrt{2}$.
[K-G] 112.

6.23 Answer: no. For example $\mu_1 = 2^{1/2}\delta_{(0,1)}$, μ_2 is the measure of linear density 1 on the set $\{(x, 1 - |x|) : x \in [-1, 1]\}$.
Remark: one can construct the examples where $\mu_1(\mathbb{R}_+^2) < \infty$, $\mu_2(\mathbb{R}_+^2) = \infty$ with μ_1, μ_2 absolutely continuous with respect to the plane Lebesgue measure.

6.24 Hint: recall the construction of the Bernstein set, see [Ox] Th. 5.3.

6.25 [H-S] 19.75, see also [G] 252.

6.26 [H-S] 10.51.

6.27 Hint: consider the balls in G centered at x/m of radii r_m, where $x \in \mathbb{Z}^n$, $r_m = n^{1/2}/m^{1 + 1/n}$.
AMM 80, 1973, 410–411, J. C. Kieffer.

6.28 [G] 508.

6.29 Hint: Approximate g and h by monotone sequences of continuous functions.
[Kl2] 5.30.

6.30 Answer: for example $f(x) = \exp(-N(x))$ where $N(x)$ denotes the number of 0 and 9 in the decimal expansion of x.
AMM 76, 1969, 427–428, 5598, S. J. Metz; B. A. Fusaro, J. G. Mauldon.

6.31 Answer: yes. Construct f as the limit of a sequence of piecewise linear functions.
AMM 72, 1965, 785–786, E1715, R. F. Jolly; G. P. Speck.

6.32 [Di] 3.15.1.

6.33 [Si] I, 166, Section 55.

6.34 [Sr] I, VI, 6.8.

6.35 Hint: take $h(x) = \Sigma_{n=1}^{\infty} 2a_n/3^n$ where $x = \Sigma_{n=1}^{\infty} a_n/2^n$, $a_n \in \{0, 1\}$.
Mathematica 9, 1935, 83–85, S. Ruziewicz; *AMM 90*, 1983, 573, 6378, Yi Hong, Jingcheng Tong; M. Renardy.

6.36 Answer: let $F_n = \{x \in [0, 1] : f(x) = f(y)$ for some $y \geqslant x + 1/n\}$ and $E = \bigcap_{n=1}^{\infty} ([0, 1] \backslash F_n)$. It is obvious that $E \in x = \sup \{y \in [0, 1] : f(y) = f(x)\}$. [H-S] 6.101; this construction is due to S. Banach.

6.37 Answer: $E = \bigcup_m \bigcap_n (r_n - 2^{-(n+m)}, r_n + 2^{-(n+m)})$, where (r_n) is an enumeration of rationals in $(0, 1)$.
[G] 451.

6.38 *AMM 91*, 1984, 520–521, 6419, W. Vervaat; K. Stromberg. 6.44 may be also useful. See 5.15.

6.39 [Kl3] 3.17.

6.40 [G] 241. See 5.15.

6.42 *AMM 68*, 1961, 72, 4879, L. J. Wallen; D. J. Newman.

6.43 [Hi2] Th. 13.1.7.

6.44 Z. Semadeni, *Banach Spaces of Continuous Functions*, PWN, Warszawa, 1971, IV.14.5.3; *AMM 85*, 1978, 582–584, R. Cignoli, J. Hounie; *AMM 87*, 1980, 816, J. Henle; *AMM 90*, 1983, 475–476, R. Boas.

6.45 Hint: consider $\mathbb{R}$ as the linear space over $\mathbb{Q}$. Observe that these functions must be discontinuous.
AMM 64, 1957, 598–599, 4706, S. W. Golomb; R. H. E. Denniston, H. Flanders.

6.46 Hint: begin with $g(x) = x$. Observe that f cannot be measurable.

6.47 [G] 453.

6.49 Answer: $f_n \to 0$ in measure, $\lim_{n\to\infty} f_n(x)$ does not exist for any x.
A supplementary question: how can one extract a subsequence of (f_n) converging a.e.?
[K-G] 168, 169; [Te] 4.29.

6.50 *AMM 93*, 1986, 133, 6509, Boo Rim Choe.

6.51 *AMM 81*, 1974, 67–68, R. E. Dressler, K. R. Stromberg.

6.52 [R-N] II.31.

6.53 [Kl2] 4.10; [R-N] II.47.

6.54 [G] 271.

6.56 [Te] 4.51. See *CMB 21*, 1978, 497–498, D. Borwein, S. Z. Ditor.

6.57 *NAW 13*, 1965, 138–139, 20, N. G. de Bruijn, J. Holbrook.

6.58 Answer: this limit exists and it is equal to J if and only if the set $\{x: f(x) \neq J\}$ can be arranged into a sequence (a_n) so that $\lim_{n \to \infty} f(a_n) = J$. *AMM 68*, 1961, 677, 4924, J. H. Blau.

6.59 *AMM 72*, 1965, 200, 5175, R. Bojanić; L. Junker, D. Ž. Djoković.

6.60 Answer: no. Consider $f = \Sigma_{n=1}^{\infty} f_n/2^n$, where $f_n(x) = (x - r_n)^{1/2}$ $\times \chi_{(r_n, r_n + 1)}$, (r_n) is an enumeration of $\mathbb{Q}$. *AMM 75*, 1968, 1025, 5546, T. S. Shores; D. Henkel.

6.61 Hint: integrate G'/G^p for $G(x) = \int_0^x |f|^p$.
Answer: only for $f \equiv 0$.
AMM 91, 1984, 589–590, 6422, H. Diamond, B. Montgomery; A. Meir.

6.62 [G] 261.

6.63 Answer: not necessarily. Consider the function equal to $(-1)^n$ on $(2^{-n-1}, 2^{-n}]$ or $f(t) = \sin(1/t)$.
AMM 67, 1960, 385–386, 4849, L. A. Rubel; B. McMillan, I. J. Schoenberg.

6.64 Answer: $f_n(x) \to x/2$.
AMM 75, 1968, 422–425, 5487, R. O. Davies; H. A. Guess, M. E. Muldoon, A. C. Zaanen.

6.65 *AMM 65*, 1958, 50, 4727, G. R. MacLane.
Remark: observe that for any positive continuous function f on $[0, \infty)$ such that $\int_0^\infty f = \infty$ there exists $h > 0$ satisfying $\Sigma_{n=1}^\infty f(nh) = \infty$.
AMM 64, 1957, 119–120, 467, K. L. Chung; A. R. Hyde, N. J. Fine.

6.66 Hint: the assumption that these functions are piecewise continuous simplifies considerably the computations.
Answer: $s_\varepsilon = -2/\log \varepsilon$.
[CHM] 4/1957, R.5.

6.67 *AMM 92*, 1985, 433–434, 6440, M. S. Klamkin, J. McGregor, A. Meir; G. A. Edgar.
For the Liapunov theorem see for example J. Diestel, J. J. Uhl, *Vector measures*, AMS, Providence, 1977.

6.68 Hint: begin with $a = 1/2$. Prove by induction on n. Then represent a in the form $\Sigma_{k=1}^\infty a_k/2^k$, $a_k \in \{0, 1\}$.
Remark: this problem is closely related to the Liapunov theorem on the convex range of nonatomic vector measure: *Izv. AN SSSR 4*, 1940, 465–478; *Bull. AMS 54*, 1948, 416–421, P. R. Halmos; [R3] Th.5.5.

6.69 Answer: the extreme points are the indicators of measurable sets. Each function in the considered set can be represented in this manner. Compare

this result with the Krein-Milman theorem for the space L^∞ with the weak-* topology.
[CHM] 9/1952, R.3.

6.70 Answer: consider for example $E = [0, 2\pi]$ and $f_n(x) = \sin nx$.

6.71 *AMM 87,* 1980, 583, 6309, M. F. Driscoll.

6.72 *NAW 13,* 1965, 238–239, 34, I. J. Schoenberg.

6.73 Remark: it is not difficult to generalize this result to the case $L^{p+1} \subseteq L^p$.
AMM 75, 1968, 553–555, 5490, J. T. Darwin; B. S. Lalli, R. Singh, P. R. Chernoff, J. A. Dyer.

6.74 See 3.20. *AMM 92,* 1985, 347–349, P. Biler.

6.75 [H-S] 20.64, 20.65; [Te] 5.117, 5.118.

6.76 [G] 273.

6.77 Answer: no. For example consider the measure μ on $\mathbb{N}$, $\mu(n) = 1/2^n$, and $f_n(x) = 2^n/n$ if $x = n$ and $f_n(x) = 0$ elsewhere. The least majorant h of f_n is not integrable.
Remark: the Doob theorem implies that the uniform integrability of the subset of $L^1(\mu)$ is equivalent to the compactness of its closure in the weak topology of $L^1(\mu)$.

6.78 [Te] 5.36, 5.37.

6.79 Hint: apply the Egorov theorem and the Hölder inequality. A well-known fact, for example *AMM 81,* 1974, 387–389, W. R. Wade.

6.80 *NAW 13,* 1965, 259–260, 60, A. C. Zaanen.

6.81 [T] 11.10.

6.82 [G] 185, 187.

6.83 Remark: the general formula for the length of the graph is
$L = \int_0^1 (1 + (f')^2)^{1/2} + V(f - \int_0^x f', [0, 1])$. *AMM 84,* 1977, 746, 6074, H. L. Montgomery, P. Borwein; *AMM 83,* 1976, 663–664, 6007, R. Sandberg; R. A. Christiansen; [R2] 8.10.

6.84 For example [H-S] 17.34; [K-G] 229; [Te] 6.32.

6.85 Answer: consider a dense open set G in $[0, 1]$ of measure $1/2$ which is disjoint union of the intervals I_n. Let J_n denote the closed interval in I_n with common center and $m(J_n) = m(I_n)^2$. Define f to be equal to 1 in the center of J_n, $0 \leqslant f \leqslant 1$ and $f = 0$ outside the union of J_n. Observe that $\overline{\int} f - \underline{\int} f \geqslant 1/2$.
AMM 84, 1977, 205–206, C. Goffman; [K11] 4.28.

6.86   **Answer:** for example $x^2 \sin(1/x^2)$.
[Te] 6.89; [T] 10.7.III.

6.89   *AMM 85*, 1978, 35–37, L. Takács (and many references therein); for the classical example see [R-N] II. 24.

6.90   **Remark:** if E is any of the sets E_k in 6.11, then $f(x) = \int_0^x (\chi_E(t) - 1/2)\,dt$ is not monotone on any interval, its derivative exists a.e. but not everywhere. [Kl2] 5.21; *AMM 81*, 1974, 349–354, Y. Katznelson, K. Stromberg; *Bull. Soc. Math. France 43*, 1915, 161–248, A. Denjoy.

6.91   *AMM 92*, 1985, 671, 6459, M. Bowron; F. S. Cater; see [R2] Th.8.10.

6.92   **Answer:** recall the properties of the Riemann function in 6.87, or consider $f(x) = \sum_{n=1}^{\infty} 2^{-n} h(x - r_n)$, where (r_n) is an enumeration of the rationals and $h(x) = x^2 \sin(1/x)$.
For example *AMM 74*, 1967, 215, 5366, H. D. Keesing; L. A. Steen.

6.93   **Hint:** consider $2\chi_{[0,1/3)} + \chi_{[1/3,2/3)} + 1.5\chi_{[2/3,1]}$.
Answer: no.
[G] 282.

6.94   [G] 126.

6.95   [G] 431.

6.96   **Hint:** show that if for the function g, $g(a) = g(b)$, $a < b$, then there exist $a < c$, $d < b$ such that $g'_+(c) \leqslant 0$, $g'_+(d) \geqslant 0$. Apply to the function $g(x) = f(x) - f(a) - (x-a)(f(b) - f(a))/(b-a)$ to   show   that $f'_+(c) \leqslant (f(b) - f(a))/(b - a) \leqslant f'_+(d)$.
For example *AMM 87*, 1980, 657–658, W. J. Knight; [T] 11.4.

6.97   [Di] 8.5.1,b.

6.99   [H-S] 18.34.

6.100   [H-S] 18.36.

6.101   [H-S] 18.41.

6.102   [G] 435, 436.

6.103   For example *AMM 69*, 1962, 312–313, E. L. Spitznagel; U. Dudley, D. A. Moran.

6.104   *AMM 91*, 1984, 307–308, F. S. Cater; see *AMM 89*, 1982, 691, P. Billingsley; *AMM 91*, 1984, 254–256, D. P. Minassian, J. W. Gaisser.

6.105   **Answer:** the Van der Waerden function is a good example here.
[CHM] 7/1960, R.8.

6.106 [H-S] 17.7; *Bull. AMS 70*, 1964, 199–213, J. P. Kahane.

6.107 **Hint:** a formal differentiation of $y(t)$ gives a divergent series.

6.108 For far-reaching generalizations see *J. Analyse Math. 51*, 1988, 62–90, M. Hata. *Bull. AMS 44*, 1938, 519, I. J. Schoenberg; [B] 107; *AMM 90*, 1983, 283, Liu Wen; *RMS 95*, 1985, 344.

6.109 *AMM 90*, 1983, 281–282, E. E. Posey, J. E. Vaughan and *Rocky Mt. J. Math. 16*, 1986, 661–668; *AMM 92*, 1985, 209–211, V. Drobot, M. Morayne. See *J. London Math. Soc. 38*, 1988, 442–452, T. W. Körner.

6.110 **Answer:** no. For example such a function does not exist for nowhere differentiable function f.
AMM 84, 1977, 143, 6027, P. Hanser, S. J. Sidney.

6.111 [R-N] I.9.

6.112 *AMM 93*, 1986, 65–66, 6463, J. M. Bayod; R. O. Garcia.

6.113 [G] 121.

6.114 [*R-N*] I.5; [H-S] 17.18.

6.115 **Hint:** approximate f by piecewise linear functions on the partition of the interval giving a good estimate of the variation of f. The Fubini theorem from the preceding problem may be useful.
Answer: consider $f(x) = x^2 \cos(x^{-b})$, $0 < x < 1$, $1 < b < 2$, for which such points do exist.
S. Saks, *Theory of Integral*, Hafner, New York, 1937, 9.3; *AMM 83*, 1976, 538–546, F. N. Huggins; *AMM 86*, 1979, 480–482, A. M. Russell; [T] 11.11.

6.118 [G] 198.

6.119 [G] 262; [Sr] XIII.6.4.

6.120 **Hint:** such a measure would be the functional on the polynomials acting so that $P(i) = \int_{-1}^{1} P(t)\,d\mu(t)$. Of course i can be replaced here by any complex number outside $[-1, 1]$. One may use also 6.117.

6.121 [G] 369.

6.122 [R-S] X.25.

6.123 [G] 286.

6.124 *AMM 72*, 1965, 89–90, 5163, S. D. Chatterji; P. R. Chernoff, N. J. Fine.

6.125 [R-S] VII.28.

6.126 [K-G] 129, 130; M. Kac, S. Ulam, *Mathematics and Logic, Retrospect and Prospects*, F. A. Praeger, New York, 1968, I, Section 17.

6.127 *NAW 18*, 1970, 311.

6.128 *AMM 89*, 1982, 707, 6337, R. D. Mauldin, J. Mycielski; A. Fieldsteel.

6.129 **Hint:** describe the sets $A_N = \{x : n(x) = N\}$.
[CHM] 6/1959, R.18.

6.130 **Hint:** it is not difficult to construct a σ-algebra containing all Borel sets which does not contain the considered product.
AMM 84, 1977, 67, 6023, S. J. Sidney; D. Ž. Djoković.

6.131 **Hint:** consider the measures m_r, $0 < r < 1$, on the product $\Pi_{n=1}^{\infty} \{0, 1\}$, $m_r = \Pi_{n=1}^{\infty} \mu_r$, where $\mu_r(\{0\}) = r$, $\mu_r(\{1\}) = 1 - r$ and transport them onto $[0, 1]$.
[H-S] 22.39.

Answers to Chapter 7
Analytic Functions

7.1 **Hint:** $\lim_{n \to \infty} n^{-1} \sum_{k=1}^{n} (k/n)^{ia} = \int_0^1 x^{ia} \, dx$.
Answer: $\{z : |z| = (1 + a^2)^{-1/2}\}$.
[Dn] IV.11.

7.2 [MS] 146, 35.

7.4 **Answer:** this series converges absolutely for $\operatorname{Re} z < 1$ to $1/(1 - z)$ and diverges in $\{z : \operatorname{Re} z > 1\} \cup \{1 + 0i\}$. For $z = 1 + it$, $t \neq 0$, convergence is not absolute.
[CHM] 7/1949, F.7.

7.5 **Hint:** $s_n = n2^{n-1}$.

7.6 **Answer:** $\{z : |\operatorname{Re} z| < 1, |\operatorname{Im} z| < 1\}$.
[MS] 143, 30.

7.7 **Answer:** the first series converges for real z only.
[As] 1.2.8.

7.8 **Hint:** $\sum_{n=1}^{\infty} a_n z^n = (1 - ze^{-z})^{-1}$ which is the analytic function in a disk of radius strictly greater than 1.
AMM 64, 1957, 279, 4684, D. J. Newman; D. Shanks.

7.9 **Answer:** $\Sigma_{n=0}^{N-1} a_n/N$.
AMM 72, 1965, 326, 5180, A. E. Livingston; J. Boersma.

7.10 **Hint:** without loss of generality, $s = 0$. In the second part apply the already obtained result and remark that $\Sigma_{n=0}^{\infty} p_n r^n/(1-r) = \Sigma_{n=0}^{\infty} (p_0 + \cdots + p_n)r^n$.
[Dn] III.13, 17.

7.11 **Hint:** (i) $\Sigma_{n=0}^{\infty} a_n r^n/(1-r) = \Sigma_{n=0}^{\infty} s_n r^n$;
(ii) $\Sigma_{n=0}^{\infty} a_n r^n/(1-r)^2 = \Sigma_{n=0}^{\infty} (s_0 + \cdots + s_n)r^n/n$, $\Sigma_{n=0}^{\infty} nr^n = r/(1-r)^2$.
[Dn] III.14, 15.

7.12 **Hint:** let $r_n = (\Sigma_{k=n}^{\infty} k|a_k|^2)^{1/2}$. Show that for $0 < r < 1$ $|f(r) - s_n| \leqslant nr_0(1-r) + r_{n+1}(1-r)^{-1/2}(n+1)^{-1/2}$.
[He] II.5.17, [Hi1] 11.5.2.

7.13 [As] 1.2.3.

7.14 [Pr] VI.6, [He] VI.2.4.

7.15 [Di] 9.3.3.

7.16 [He] II.5.9.

7.17 **Hint:** the series converges to an analytic function in any connected region of the complex plane where $|ze^{-z}| < e^{-1}$.
[P-S] I.3.260.

7.18 **Hint:** $(S_n(z))$ is a normal family of analytic functions on the complement of an f, f^{-1} invariant set. The series converges for z satisfying $\lim_{n \to \infty} S_n(z) = \infty$.
AMM 85, 1978, 290–291, 6109, S. P. Lloyd; I. N. Baker; *J. Math. Pures Appl.*
41, 1962, 339–351, P. J. Myrberg.

7.20 [Hi1] 8.8.2.

7.21 [Le] 5.7.

7.22 *AMM 93*, 1986, 305–306, 6481, I. J. Schoenberg; W. Janous.
It is a modification of a result of G. Szegö, *Coll. Papers*, vol. 2, 171.

7.23 [Hi1] 8.3.4.

7.24 [He] V.3.11.

7.26 **Hint:** begin with the construction of a disk centered in the vertex a_j of the polygon conv$(a_1, \ldots, a_n)$ of radius $2Hm/n$, where m is the multiplicity of a_j in $a_1, \ldots, a_n$.
[Le] 4.1.

7.27 Compare [Dn] II.19 and [He] IV.4.12.

7.28 [He] V.3.29. See also *AMM 74*, 1967, 854–855, Ch. Fefferman and *AMM 88*, 1981, 253–256, J. L. Brenner, R. C. Lyndon.

7.29 **Hint:** apply the Cauchy integral formula for the function $f(z)\overline{f(\bar{z})}$ on the semicircle of the diameter $[-1, 1]$.
[He] V.3.30.

7.30 **Answer:** $f(z) = k\exp(z/2k)$, $f(z) = c\sin(z/c)$, $f(z) = z$.
[Hi1] 10.7.3; *AMM 68*, 1961, 511, 4910, J. S. Frame; J. W. Ellis.

7.31 **Hint:** apply the method of indetermined coefficients.
[He] VI.2.13.

7.32 [M] I.5. Compare related topics in: *AMM 76*, 1969, 315–316, 5586, H. D. Keesing; D. A. Herrero; *AMM 85*, 1978, 246–256, J. D. Gray, S. A. Morris.

7.33 *Uspekhi Mat. Nauk 15*:1, 1960, 191–194, V. K. Dzjadyk; *AMM 71*, 1964, 265–267, A. W. Goodman.

7.34 **Hint:** suppose first that $u, v \in C^2$. Then complete the proof recalling the Radó theorem, see [R2] 12.13.
AMM 86, 1979, 490–491, B. Walsh; *AMM 85*, 1978, 333–337, H. S. Bear, G. N. Hile.

7.35 [As] 3.2.15.

7.36 **Answer:** 0.
AMM 75, 1968, 800, 5519, W. O. Egerland; D. A. Hejhal.

7.37 [As] 3.1.10, 3.1.5.

7.38 [As] 2.2.7.

7.40 **Answer:** (i) e.g. $(z - a)(z - b)$ on $\mathbb{C}\backslash[a, b]$.
[As] 2.2.1, 2.

7.41 [As] 2.3.1.

7.42 [Co] VI.2.2.

7.43 [Ho] 3.4.4.

7.44 [As] 3.4.1.

7.45 **Hint:** begin with the case $f(0) = 0$. Consider the function $g(z) = f(Rz)(2A(R) - f(Rz))^{-1}$ for $|z| < 1$.
[Co] VI.2.4; [T] 5.5.

7.46 Hint: apply the Schwarz lemma to the function $M(f(z) - 1)/(M^2 - f(z))$. Then apply this result to h' and use the Rouché theorem. [Di] 10.3.5.

7.47 [He] V.9.8.

7.49 [Pr] 7.2.2.

7.50 [Pr] 7.2.2; [S1] 1.59. See *AMM 84*, 1977, 368–370, D. J. Uherka, A. M. Sergott; *Proc. AMS 100*, 1987, 390–392, G. Harris, C. Martin.

7.51 Hint: denote by a_r, b_r real numbers in $[0, 2\pi)$ such that $\max_{s,t \in \mathbb{R}} \{|f(re^{is}) - f(re^{it})|\} = |f(r \exp(ia_r)) - f(r \exp(ib_r))|$. One obtains the result applying the Schwarz lemma to the function $f(z \exp(ia_r)) - f(z \exp(ib_r))$.
[CHM] 9/1953, F.1.

7.52 Hint: for $|z| < r$

$$|F(z)|^n \leqslant \frac{1}{2\pi(r - |z|)} \int_0^{2\pi} |F(re^{iu})|^n \, du$$

AMM 90, 1983, 339–340, 6362, F. S. Cater.

7.53 Remark: it suffices that $\operatorname{Re} A \geqslant (\sqrt{5} - 1)/2$.
AMM 84, 1977, 223–224, 6033, S. Miller; R. Vermes.

7.54 [Co] VII.4.3.

7.55 [Co] VII.4.4.

7.56 [Co] VII.4.6.

7.57 [Hi1] Th. 18.5.1, 2.

7.58 Answer: the Picard theorem implies that such a function is linear. Indeed, it takes on each real value at most once. If $f = u + iv$ satisfies the hypothesis then $\partial u/\partial x = \partial v/\partial y$ is of constant sign on the real axis since v is of constant sign in the upper half-plane.
For example *AMM 67*, 1960, 597, 4857, I. N. Baker, D. J. Newman.

7.59 For example [As] 3.4.2, [R2] 12.6.

7.60 [As] 3.4.4.

7.62 [As] 3.4.3.

7.63 [Co] XII.4.1.

7.66 [As] 3.1.8.

7.67 **Hint:** define $g(e^z) = f(z)$. Show that g is analytic on $\mathbb{C}\backslash\{0\}$ and its kth Laurent coefficient is equal to $(2\pi)^{-1} \int_0^{2\pi} f(x + iy)e^{-k(x+iy)} dy$, $x \in \mathbb{R}$. Estimate these coefficients for large $|k|$.
[He] VI.1.15.

7.68 **Hint:** it is well known that the image of the unit disk by $z/(1 - z)^2$ is equal to $\mathbb{C}\backslash(-\infty, -1/4]$.
Answer: for example $B(z) = -z$.
AMM 91, 1984, 653, 6425, J. Miles, L. Rubel.

7.69 **Hint:** reduce the question to the problem of conformal equivalence of the annuli. The mapping $w(z) = (1 + (1 - \varepsilon^2)^{1/2})z/2\varepsilon + (1 + (1 - \varepsilon^2)^{1/2})/2\varepsilon z$ may be of great use.
AMM 84, 1977, 396, 6047, C. D. Minda; P. R. Chernoff; [R2] 14.22.

7.70 **Hint:** if F has a zero, then F is a polynomial (otherwise ∞ is an essential singularity). If F does not vanish then $F = \exp g$ for some entire function g. Show that g is linear.
AMM 90, 1983, 488–489, 6279, L. A. Rubel; P. R. Chernoff, J. Essick, J. Maly, Z. Vlasek.

7.71 [Pr] VIII.2.

7.72 **Hint:** the Picard theorem implies that the range of $f(z) - z$ contains either 0 or a period of f.
AMM 66, 1959, 315, E1334, B. Randol; A. Rosenthal.

7.73 **Hint:** represent f, g as $f(z) = e^{F(z)} + z$, $g(z) = e^{G(z)} + z$. The equation $f \circ g(z) = z$ is equivalent to $\exp(F(e^{G(z)} + z)) = -e^{G(z)}$ or $F(e^{G(z)} + z) = G(z) + (2n + 1)i\pi$. Apply the Picard theorem.
AMM 64, 1957, 682, 4717, P. Gallagher; P. Henrici.

7.74 **Hint:** if there would be two such points, then the iterates of f form a normal family in the sense of Montel. Therefore these iterates are uniformly bounded on a neighborhood of the origin $(f(0) = 0)$, so $|a_1^n| \leqslant \text{const}$ for all n from the Cauchy inequality.
AMM 68, 1961, 512, 4911, P. Ungar; J. H. van Lint.

7.75 [Co] V.3.9.

7.76 **Hint:** for $|z| = 1$, $z \neq 1$, and $P(1) > 0$ $|(1 - z)P(z)| > a_n - [(a_n - a_{n-1}) + \cdots + (a_1 - a_0) + a_0] = 0$.
[Dn] II.17.

7.77 **Hint:** observe that $f(rz) - f(z) = rzf(z)$.
AMM 72, 1965, 431, 5193, L. Batson; N. J. Fine.

7.78 *C. R. Acad. Sci. Paris 250*, 1960, 1953–1954, M. Parodi.

7.79 *AMM 72*, 1965, 562–563, 5207, H. S. Shapiro, D. J. Newman; P. M. Gibson.

7.80 [VLA] 799.

7.81 **Hint:** first show that $(2\pi)^{-1} \int_0^{2\pi} |f(e^{it})| \, dt \leqslant 1 + |a_1|^2/4$. If f does not vanish then $f(z)^{1/2} = 1 + b_1 z + b_2 z^2 + \cdots$, where $b_1 = a_1/2$. On the other hand $1 + |b_1|^2 + |b_2|^2 + \cdots < 1 + |a_1|^2/4$, a contradiction.
[CHM] 9/1958, F.4.

7.82 *AMM 74*, 1967, 452–453, 5389, D. J. Newman; R. Breusch, M. D. Mavinkurve.

7.83 **Hint:** apply the Rouché theorem to the functions $f(z)$, $g(z) = cz - f(z)$, where $c = \inf_{zebdG} \operatorname{Re} \bar{z}f(z)/\sup_{zebdG} |z|^2$.
AMM 89, 1982, 210–211, Mau-Hsiang Shih.

7.84 **Hint:** from the real mean value theorem

$$\int_a^b f(r)|g(z)| \, |dz| = |g(z_0)| \int_a^b f(r) \, |dz|$$

[Dn] VII.55.IV.

7.85 **Hint:** use the Rouché theorem.
AMM 80, 1973, 45–46, Å. Samuelson.

7.86 **Hint:** apply the argument principle. Observe that C has the continuous tangent.
[Hi] 16.1.7.

7.87 **Answer:** this equation has for every $a > 0$ infinitely many real roots and for every $0 < a < 1$ two imaginary roots also.
[Hi1] 9.2.7.

7.88 **Answer:** countably many roots.
Hint: it may be verified directly separating the real and the imaginary parts of the equation. Another method: apply the Borel theorem ([S-Z] VII.11.2) to the function $((\sin \sqrt{z})/\sqrt{z} - 1)/z$ of fractional order. The corollary of this theorem gives the asymptotics of the roots (z_n): for all $a > 1$ $\sum_{n=1}^{\infty} |z_n|^{-a} < \infty$.

7.89 **Hint:** consider the function $g(z) = f(z) \prod_{k=1}^{m} (1 - z/z_k)^{-1}$, where $z_1, \ldots, z_m$ denote the zeros of f.
[Co] XI.1.2, see [Co] VI.1.5.

7.90 **Hint:** apply the residue formula.
[Co] XI.1.4.

7.91 **Hint:** $|f(p/q + iy) + ie^{-q^2}/y| < q \sum_{n=1}^{\infty} n^2 e^{-n^2}$.
[Hi1] 10.4.5.

7.92 **Answer:** for example $e^z + z$.
[Pr] VIII.4, [Dn] IV.6.

7.93 [Hi1] 11.3.8.

7.94 **Answer:** the Lindelöf-Malmquist function

$$L_b(z) = \sum_{n=0}^{\infty} \left(\frac{z}{\log(n + b)} \right)^n, b > 1$$

converges to zero along all half-lines but $\mathbb{R}^+$. The function $\exp(-L_b(z)) - \exp(-L_c(z))$, where $b, c > 1$, $b \neq c$, satisfies the conditions of the problem.
[Dn] X.87, [Hi1] 11.3.6, 7.
Another example:

$$f(z) = \frac{1}{2\pi i} \oint_C \exp(e^w) \exp(-\exp(e^w))(w - z)^{-1} \, dw$$

where $C = \{(t, -i\pi/2) : t \geq 0\} \cup \{(0, s) : -i\pi/2 \leq s \leq 3i\pi/2\} \cup \{(t, 3i\pi/2) : t \geq 0\}$. *Acta Math. 42*, 1920, 285–308, G. Mittag-Leffler; *Math. Ann. 91*, 1924, 316–320, K. Grandjot; *AMM 83*, 1976, 192–193, D. J. Newman.

7.95 **Hint:** compare with the preceding problem.
AMM 67, 1960, 929, 4881, B. Randol; J. C. Kneser.

7.96 **Hint:** observe that $|a_n| \leq (eb/n)^{n/b}$.
[Di] 9.11.3.

7.97 [He] VI.2.12.

7.98 **Hint:** construct a sequence of positive integers (k_n) such that $k_1 = 1$, $(n/(n - 1))^{k_n} > F(n + 1)$ for $n > 1$ and consider $f(z) = 1 + \sum_{n=2}^{\infty} (z/(n - 1))^{k_n}$.
[Di] 9.14.4.

7.99 [He] VI.2.8.

7.100 **Hint:** an elementary proof uses the Borel-Carathéodory inequality (see 7.45) and the inequality $M(r) \leq 3A(2r)$ valid for all large r.
AMM 85, 1978, 265–268, L. Zalcman. See also *AMM 88*, 1981, 45–47, D. S. Bridges; *AMM 89*, 1982, 239, A. S. Madguerova.

7.102 **Hint:** the Zhukovskii map $w(z) = (z + 1/z)/2$ may be useful.
[He] VI.4.30.

7.103 [Ev] 3.7.1.

7.105 [R-S] Th. XII.17, XII.29.

7.106 Hint: apply the Carlson theorem to $(f_2 - f_1)/\sin \pi z$, where f_1, f_2 satisfy the hypotheses of the problem.
[R-S] XIII.109.
  For an extension of the Carlson theorem related to this problem see *J. Math. Anal. Appl. 129*, 1988, 131–133, R. P. Boas, A. M. Trembinska.

7.107 Hint: first prove the following convergence criterion:
  If $\lim_{n \to \infty} n^{1/2} b_n = 0$, $\Sigma_{n=1}^{\infty} n^{1/2} |b_n - b_{n-1}| < \infty$, $\sup_n |\Sigma_{k=1}^{n} a_k / n^{-1/2}| < \infty$,
then the series $\Sigma_{n=1}^{\infty} a_n b_n$ converges. Apply this for $|z| = 1$, $z \neq 1$.
[VLA] 449, 90.

7.108 [Dn] V.13, 14; [As] 1.2.8.

7.109 Hint: $|a_n| \leqslant \mathrm{Re}\, a_n / \cos c$ and the terms of the series may be rearranged.
[Dn] VII.57.II.
Remark: the statement of Ex. 5.7.1 in [Hi1] is not correct.

7.110 Hint: if $f_m(z) = \Sigma_{n=m}^{\infty} a^n z^{b^n}$ then $f_m(z \exp(2\pi i k b^{-n})) = f_m(z)$ for positive integers k, hence f has a dense set of singularities on the circle. [VLA] 1084; [Dn] VII.57.II.
  For (a generalization of) the Hadamard theorem on lacunary series see [T] 7.4.3, 4 and [Hi1] 11.7.1, 2 (the Fabry theorem).

7.111 [Hi1] 5.7.5.

7.112 [Dn] X.82.I.

7.113 [Dn] X.83.II.

7.114 [Hi1] 11.7.2.

7.115 The P. S. Urysohn example: [Hi1] 10.4.6.

7.116 [Hi1] 11.7.4.

7.117 Hint: take a lacunary subseries $H(z) = \Sigma_{k=1}^{\infty} a_{n_k} z^{n_k}$ such that $\limsup_{k \to \infty} |a_{n_k}|^{1/n_k} = 1$ and $n_{k+1} > 2 n_k$ and put $f_0(z) = \Sigma_{n=1}^{\infty} a_n z^n - H(z)$. Now consider the whole class of series $H(z) = \varepsilon_1 f_1(z) + \varepsilon_2 f_2(z) + \cdots$, where $\varepsilon_n \in \{-1, 1\}$ and f_n contain different powers. There are continuum choices of signs and only countably many roots of unity; hence if all these series would be continuable, there would be two of them continuable in the direction of the same root of unity. Their difference would also be continuable but it does not contain f_0 and it is a lacunary series without continuation.
[Dn] XI.94.III.

7.118 Answer: two such series are in the same class if and only if there exists an integer N such that their terms coincide for $n \geqslant N$.
[Hi1] 10.3.3.

7.119 [Dn] XIII.111.

7.120 [Ev] 3.7.4.

7.121 **Hint:** the cases $D = \mathbb{C} \cup \{\infty\}$ or $\bar{\mathbb{C}} \backslash D$ countable are simple. In the general situation one may assume that $\infty \in D$, so the boundary of D is bounded. Choose a countable dense set of points (b_n) in the boundary and a sequence (a_n) so that $a_n > 2\Sigma_{k=n+1}^{\infty} a_k$ (for example $a_n = a^{-n}$, $a > 3$). Put $f(z) = \Sigma_{n=1}^{\infty} a_n(z - b_n)^{-1}$.
[Hi1] 10.4.2, see another construction in [N2] Th.6.3.3.

7.122 **Answer:** choose f, g so that $h \equiv 0$.
For an example to the second part see [Hi1] Th. 11.6.1. Another version of the Hadamard theorem: [Dn] X.88.I.

7.123 [Dn] X.88.II.

7.124 [Dn] X.88.III.

7.125 **Hint:** it suffices to show that there exists an integer $k > 1$ such that $\lim_{z \to 1} \Sigma_{n=1}^{\infty} a_n^k z^n = \infty$ and then apply the Hadamard theorem. The following fact may be useful: There exists a polynomial P such that $P(0) = 0$, Re $P(z) > 1$ in the considered domain.
AMM 68, 1961, 186–187, 4896, H. S. Shapiro; D. J. Newman.

7.126 **Hint:** the image of the circle $|z| = r < 1$ is a closed Jordan curve with exactly two points on the real axis corresponding to $z = \pm r$. Show that $|c_n| r^n \leqslant nr$ which follows from the estimate of the Fourier coefficients of the odd function Im $f \geqslant 0$ in $(0, \pi)$: $|b_n| \leqslant nb_1$.
[Di] 17.4.9.
Remark: the conclusion remains true without any restriction on (c_n). This famous Bieberbach conjecture was proved in 1984 by L. de Branges, see *AMM 93*, 1986, 505–514, J. Korevaar; *Math. Intell. 7/2*, 1985, 23–25, 32,Ch. Pommerenke; *ibidem 8/1*, 1986, 40–47, O. M. Fomenko, G. V. Kuz'mina.

7.127 [R4] 8.4.1.

7.128 **Hint:** the great Picard theorem may be useful.
AMM 85, 1978, 505–506, 6117, M. J. Pelling, L. A. Rubel; [N2] 6.4.2.

7.129 **Hint:** use the small Picard theorem.
Remark: $e^z + e^{z+i\pi} + e^0 \equiv 1$.
AMM 85, 1978, 506, 6118, M. J. Pelling; Chen-Han Sung. See [N2] 6.4.4 and [S] I.IV.20.

7.130 [Hi2] Th. 8.2.1, [Hi1] Th. 15.3.1.

7.131 *AMM 90*, 1983, 391–394, K. Davidson.

7.132 **Hint:** for $n = 1$ apply the Mittag-Leffler theorem on the representation of meromorphic functions.
Remark: for $n > 1$ this fact is not true.
AMM 92, 1985, 365–366, 6437, B. Bagchi; L. A. Rubel.

7.133 **Hint:** first prove that if functions g_n analytic in the disk converge in L^2 on its boundary, then they converge uniformly on compact subsets of the disk. To get it apply the Cauchy integral formula.
[Le] 7.1.

7.134 [Le] 7.2.

7.135 [R3] 3.27.

7.136 [Le] 7.3.

7.137 [Le] 7.5, see for example [R2] 15.26, [Kl3] 10.11.

7.138 **Hint:** arrange all polynomials with rational coefficients in a sequence (P_n). Consider $f = \Sigma_{n=1}^{\infty} I^{k_n} P_n$, where k_n are sufficiently large integers and I is the operator of integration $I(P)(z) = \int_0^z P(w)\, dw$.
AMM 90, 1983, 331–332, Ch. Blair, L. A. Rubel.

7.139 *AMM 84*, 1977, 663, 6071, J. G. Milcetich; D. D. Brannan.

7.140 *AMM 85*, 1978, 55–56, 6091, H. S. Shapiro; J. B. Wilker, R. W. K. Odoni.

7.141 [S] II.9.4; [R4] 4.4.4.

7.142 **Hint:** for $n = 1$ see 7.49, then prove by induction. It suffices to show that if $(\partial F_j / \partial z_k)(z) \neq 0$ for some j, k, then the Jacobian of F does not vanish at z.
AMM 89, 1982, 587–588, J.-P. Rosay.
Remark: classical proofs of this important fact are much more complicated, see S. Bochner, W. T. Martin, *Several Complex Variables*, Princeton, 1948; R. Narasimhan, *Several Complex Variables*, Chicago, 1971.

7.143 [R-S] IX.22.

7.144 [R3] 7.14.

7.145 **Hint:** fix $z = x + iy$ and consider the function $g_s(c) = (1 - isc)^{-N-1} \times \exp(ir|y|^c) f(x + cy)$, $c \in \mathbb{C}$, $s > 0$. From the maximum principle $|g_s(i)| < 1$. Pass with s to zero.
[R3] 7.15.

7.146 **Answer:** for example $n = 2$, $F(z_1, z_2) = (z_1 + f(z_2), z_2)$ where f is an arbitrary entire function. Then $F^{-1}(w_1, w_2) = (w_1 - f(w_2), w_2)$. [S] II.4.11.

7.147 [S] II.4.11.

7.148 [R4] 2.1.1.

7.149 **Hint:** apply the result of the preceding problem.
Answer: the example is $G_1 = G_2 = \{(z, w) \in \mathbb{C}^2 : |zw| < 1\}$. Consider $F(z, w) = (zh(zw), w/h(zw))$, where h is an analytic nonvanishing function of one variable on the unit disk.
[R4] 2.1.3, 4.

7.150 [S] I.V.9, II.I.20.

7.151 **Hint:** consider the integrals

$$\oint_{|\zeta| = r} \frac{\zeta^k \, \partial F(z_1, \ldots, z_{n-1}, \zeta)/\partial z_n}{F(z_1, \ldots, z_{n-1}, \zeta)} \, d\zeta$$

where $k = 0, 1, 2, \ldots$ and $r = C(1 + |(z_1, \ldots, z_{n-1})|^m)$. Recall the proof of the Weierstrass preparation theorem.
[S] II.II.19.

Answers to Chapter 8
Fourier Series

8.1 **Hint:** apply the Abel transform.
[E] 1.2; for a geometric interpretation see *AMM 64*, 1957, 47–48, E1220, M. Goldberg.

8.2 [E] 1.3.

8.3 [E] 1.4. see 8.54.

8.4 [T] 1.1.3.1; [E] 1.5; [Hi1] Th. 5.4.5, Cor. 5.4.9; see [Ge] 210.
Remark: the condition $\lim_{n \to \infty} nc_n = 0$ in the second part is equivalent to the continuity on $\mathbb{R}$ of the sum of the series $\Sigma_{n=1}^{\infty} c_n \sin nx$. Compare with the example in 8.50.

8.6 [E] 1.7.

8.7 [E] 1.8.

8.8 **Hint:** deduce it from the preceding exercise.
[E] 1.9.

8.9 [E] 1.11.

8.10 **Hint:** suppose that $F \subseteq [-r, r]$, $r \in \mathbb{N}$. Consider $f(x) = (2N + 1)^{-1}$ $(\Sigma_{|n| \leqslant N} e^{inx})(\Sigma_{|n| \leqslant N+r} e^{inx})$ with large N.
[E] 2.19.

8.11 **Hint:** observe that $f(x_1) = f(x_3) = \cdots = f(x_{2n-1}) = 1$, $f(x_2) = f(x_4) = \cdots = f(x_{2n}) = -1$ (or with reversed signs). Consider $g(x) = f(x) - f(x + 2\pi/n)$. Count the zeros of g (with their multiplicities).
AMM 89, 1982, 454, H. Flanders, Putnam Competition 1962, PM6.

8.12 **Hint:** one may assume that $\max_{0 \leqslant x \leqslant 2\pi} |f(x)| = c_0 + \cdots + c_n = 1$. Now consider $f(x) = (1 + \cos x + \cdots + \cos nx)/(n + 1)$.
For the second estimate observe that $\pi/2 \; \max_{\pi/2 \leqslant x \leqslant \pi} |f(x)| \geqslant |\int_{\pi/2}^{\pi} f|$.
[CHM] 3/1953, S. 24.

8.13 **Answer:** the assumption implies $\Sigma_{k=1}^{n} k a_k \sin kx \geqslant 0$ on $[0, \pi]$, hence the harmonic function $h(r, x) = \Sigma_{k=1}^{n} k a_k r^k \sin kx$ is nonnegative in the upper semicircle $0 \leqslant r \leqslant 1$, $0 \leqslant x \leqslant \pi$ (from the maximum principle). Then $g(x) = \int_0^1 h(r, x) r^{-1} \, dr > 0$ for $0 < x < \pi$.
[CHM] 9/1956, S.26.

8.14 **Hint:** consider the auxiliary functions $f(x) = \Sigma_{n=1}^{\infty} n^{-2} \sin n \, \sin nx$, $g(x) = \Sigma_{n=1}^{\infty} n^{-1} \sin nx$, and apply the Parseval identity.
Note well $\int_0^{\infty} x^{-1} \sin x \, dx = \int_0^{\infty} (x^{-1} \sin x)^2 \, dx = \pi/2$.
For example *AMM 87*, 1980, 496–498, 6241, R. Baillie; E. T. H. Wang.

8.15 **Answer:** this series is conditionally convergent.
[Br] I.30. See 8.46.

8.17 **Hint:** the Fourier coefficients of $\exp(iAx)$ are equal to $\sin \pi(A - n)/(\pi(A - n))$. From the Parseval identity $\Sigma_{n \in \mathbb{Z}} |\sin \pi(A - n)/(\pi(A - n))|^2 = 1$ and for $A = t/\pi$, t complex, the formula follows.
For example *AMM 86*, 1979, 296–297, R. M. Young.

8.18 [K12] 4.11.

8.19 [E] 2.16, 6.2.5, 8.2, 10.5 and [G] 480 (partially).

8.20 [Z] VII.6.11.

8.21 **Hint:** approximate functions from L^1 by functions in L^2 and apply the Bessel inequality.
[H-R] Th. 15.

8.22 [T] 13.8.1, 13.20; [E] 2.14.

8.24 **Answer:** no. Consider $E_{p,q} = ((q - 1/2)/(2p)!, q/(2p)!)$, $p = 1, 2, \ldots$, $q = 1, \ldots, 2p$, $E = \bigcup_p \bigcup_q E_{p,q}$, so $\lim_{n \to \infty} n C_n = \infty$.
AMM 65, 1958, 453–454, 4749, J. Lehner; F. Herzog.

8.25 [E] 2.13.

8.26 Answer: no. Develop in the cosine series a function f equal 1 on $[-b, b]$, $a < b < \pi$, but $f \not\equiv 1$.
AMM 69, 1962, 816–817, 4973, A. Wilansky; N. J. Fine.

8.27 Answer: for example by $[0, c]$, $0 < c < 2\pi$.

8.28 [E] 2.12.

8.30 Hint: use the identity $s_n f = \sigma_{n+1} f + (n + 1)^{-1} \Sigma_{k=1}^n k(a_k \cos kx + b_k \sin kx)$.

8.31 For example [E] 2.8; Y. Katznelson, *An Introduction to Harmonic Analysis*, John Wiley & Sons, New York, 1968, 4.6.

8.34 A trivial (but useful) corollary: If f (integrable) and $\hat{f}$ are both of compact support in $\mathbb{R}$ then $f \equiv 0$.
For example [H-R] Th. 36.

8.35 [E] 2.7; [Ge] 190.

8.37 For example [T] 13.4.3.

8.38 Hint: assume without loss of generality that $x_0 = 0$, $f(x_0) = 0$. Apply the Riemann-Lebesgue lemma to the function $f(x)/(e^{ix} - 1)$.
AMM 87, 1980, 399–400, P. R. Chernoff; see *Proc. AMS 19*, 1968, 145, I. Richard.

8.39 Hint: dualize the statement or consider the action of T on functions with analytic continuation.

8.40 For example [E] 6.12.

8.41 [E] 3.14.

8.42 Hint: compare $s_n f$ and $\sigma_n f$. [Br] IV.2; [Ka] II.1.

8.43 Hint: let $h_\varepsilon(t) = \max(0, 1 - |t|/\varepsilon)$ for $|t| \leqslant \pi$. Then $\hat{h}_\varepsilon(n) = 2 \sin^2(\varepsilon n/2)/(\pi\varepsilon n^2) \geqslant 0$ and $f(t) = \int_0^\pi h_\varepsilon(t) \, d\mu(\varepsilon)$ for a suitable finite positive measure μ on $(0, \pi]$. Hence $\hat{f}(n) \geqslant 0$ and the result of the preceding problem applies.
[Ka] II.1.

8.44 *NAW 20*, 1972, 185–186, 298, W. A. J. Luxemburg; P. van der Steen.

8.46 [E] 7.3.3.

8.47 [T] 13.17, 13.7.1.

8.48 Hint: take small a and b/a about 1.
[T] 13.18.

8.49 **Hint:** $\int_{-\pi}^{\pi} f(x)\sin(4^n x)\,dx = \pi/4^n$.
[T] 13.19; [Ge] 215.

8.50 **Hint:** let $g(x) = \sum_{n=1}^{\infty} \sin nx/n = (\pi - x)/2, 0 < x < 2\pi$. Then $(f - g)(x)$ $= -\sum_{n=1}^{\infty} \sin nx/(n^4 + 1)n, (f - g)^{(4)}(x) = -f(x)$.
[MS] 160, 56.

8.51 **Hint:** consider the values of these functions at $x_k = 2^{k+1}\pi/15$.
Answer: all these functions are unbounded. All these properties remain unchanged if n^{-1} is replaced by a decreasing sequence (d_n) such that $\sum_{n=1}^{\infty} d_n = \infty$.
AMM 91, 1984, 259–260, 6299, R. L. Cooke, D. Borwein, P. Borwein.

8.53 **Answer:** if f is of class C^1, then the cluster points of (G_n) coincide with the graph of f. In the general case one has also some vertical segments in the points of discontinuity of f.
[K-G] 661; see [E] 10.8 and *Arch. Hist. Exact Sci. 21*, 1979, 129–160, E. Hewitt, R. E. Hewitt; *CMB 19*, 1976, 209–211, V. Swaminathan.

8.54 [L] 7.20.

8.55 **Hint:** (i) f may be represented as $f(n) = \sum_{k=1}^{p} a_k \cos(2nk\pi/p)$, where $a_k = p^{-1} \sum_{m=1}^{p} f(m) \cos(2mk\pi/p)$ and p is the period of f.
AMM 65, 1958, 349–351, H. J. Fletcher, where some interesting applications are given. See 8.56.

8.56 **Hint:** study the functions $g(z) = \sum_{r=0}^{n-1} b_r e^{2\pi i r z/n}$, $f(z) = e^{-i\pi z} g(z)/\sin(\pi z)$. Compute (using the residue theorem) the integral $(2\pi i)^{-1} \oint_{C_N} \pi f(z) z^{-1}\,dz$ along the perimeter of the rectangle with sides of length $2N + 1$, $N^{1/2}$, $N \in \mathbb{N}$, centered at the origin.
AMM 86, 1979, 786–788, E2719, a generalization of B. C. Berndt; see applications to the summation of certain series.

8.57 **Hint:** an analogy with the formula $(\lambda_{n1} + \cdots + \lambda_{nn})/n = a_0 = (2\pi)^{-1} \int_0^{2\pi} f(x)\,dx$ is striking.
[P] XIII.13.

8.58 **Hint:** it suffices to prove that $(s_{n_k} - \sigma_{n_k}) \to 0$ a.e. and apply the Fejér-Lebesgue theorem. Observe that $\sum_{k=1}^{\infty} \|s_{n_k} - \sigma_{n_k}\|_2^2 < \infty$.

8.59 **Hint:** compare the partial sums of the series and its Fejér sums.
See the example [T] 13.16.

8.60 **Hint:** the Hardy inequality $\sum_{n=0}^{\infty} |a_n|/(n + 1) \leqslant \pi \int_0^{2\pi} |f|$ would be useful.
AMM 91, 1984, 263–264, 6406, J. M. Steele.

8.61 [Br] I.34.

8.62 [E] 3.11.

8.63 **Hint:** verify this formula for trigonometric polynomials.
[E] 3.12.

8.64 [R3] 2.6.

8.66 [E] 3.10.

8.67 *AMM 85*, 1978, 832–833, M. Barr, J. G. Mauldon.

8.68 [Hi1] 9.1.7, 8.

8.69 [A] 1.2.8; see the original paper *Ann. Soc. Math. Polon. 25*, 1952, 279–287, A. Rényi.

8.70 [A] 1.3.5.

8.71 [A] 1.3.5.

8.72 **Hint:** observe that $s_n f(x_0) - f(x_0) = \int_a^b (f(t) - f(x_0))w(t) \, \sum_{k=0}^n p_k(t) \, p_k(x_0) \, dt$ and apply the Christoffel-Darboux formula from the preceding problem.
[A] 1.3.6; see [E] 5.9.

8.73 [A] 1.3.6. See the original paper *C. R. Acad. Sci. 200*, 1935, 2052–2053, J. Favard.

8.74 [A] 1.6.1, 1.6.2.

8.75 **Hint:** $\int_0^1 r_n r_1 r_2 = 0$. Apply the result on convergence of Haar series from the preceding problem. Of course there are other direct proofs of this fact.
[A] 1.7.1.

8.76 **Hint:** the completeness follows easily from the Rényi theorem in 8.69.
[A] 1.7.6.

8.77 [A] 2.3.2.

8.78 **Hint:** the Kronecker lemma may be useful: If the sequence (b_n) increases to infinity, then the convergence of the series $\sum_{n=1}^\infty a_n/b_n$ implies the estimate $s_n = o(b_n)$ for the partial sums s_n of the series $\sum_{n=1}^\infty a_n$.
[A] 2.3.3, 2.2.2.

8.79 [A] 2.11.1. See the original paper *Uspekhi Mat. Nauk 15: 5*, 1960, 77–141, A. A. Talalian (translated in *Russian Math. Surveys*).
Remark: Pál and Fekete gave examples of universal power series with the similar property of representation of every continuous function as the limit of a subsequence of partial sums; see 4.120.

8.80   Answer: yes. An interpretation: the functions $\cos(nt_j\pi)$, $j = 0, \dots, k$, are linearly independent over reals modulo the sequences converging to zero.
AMM 86, 1979, 310, E2699, E. Haddad, P. Johnson; O. P. Lossers.

8.81   Answer: yes.
Hint: each linear combination of these functions can be continued analytically for $t \in \mathbb{C}$. The absolute value of such a nontrivial expression rapidly increases for $|t| \to \infty$. Another idea of the proof in a slightly generalized situation for the functions $\exp(ir(t - t_j))$, $r \in \mathbb{C}$: represent the linear combination of these functions as the convolution of $\exp(ir\cos t)$ with the measure $\Sigma_{j=1}^{k} a_j\delta_{t_j}$. Pass to the Fourier transforms and show that the transform of $\exp(ir\cos t)$ has few zeros.
AMM 87, 1980, 762–763, 6253, M. Machover; J. F. Clarke, R. B. Israel.

8.82   Hint: consider $T^{-1}\int_0^T s(t)\,dt$ and its limit or (a proof by induction) consider $s(t) - s(t + 2\pi/\lambda_1)$.
Answer: if λ_j are different, then the necessary and sufficient condition is $a_j\lambda_j = 0$ for all j.
AMM 92, 1985, 292–293, 6442, V. Naroditsky; P. Biler.

8.83   Hint: consider the integral
$J(a) = \int_{-a}^{a}(a - |x|)f(x)\,dx = (2\pi)^{-1}a^2 \int_{\mathbb{R}}\int_{\mathbb{R}}(2\sin(at/2)/at))^2 f(x)\cos tx\,dt dx$.
Verify that $\int_{-1}^{1} f \geqslant J(1)$, $\int_{-k}^{k} f \leqslant J(k+1) - J(k)$.
AMM 84, 1977, 746–747, 6075, H. L. Montgomery; O. P. Lossers.

8.84   For example [G] 361, where a fragment of the proof of famous Wiener Tauberian theorem is given.

8.85   [R3] 9.7.

8.86   *AMM 85*, 1978, 689–690, 6131, L. A. Rubel; A. S. Goldstein.

8.87   Hint: compute for $n = 1$ the Fourier transforms of $\exp(x^2/2)(d/dx)^m$ $\times \exp(-x^2)$, $m = 0, 1, 2, \dots$.
For example [R3] 12.23.

8.88   Hint: remember the eigenvalues of $\mathscr{F}$; $f(\mathscr{F})g$ is a linear combination of $g(x)$, $g(-x)$, $\hat{g}(x)$, $\hat{g}(-x)$.
For example [K-G] 805.

8.89   Hint: the Parseval identity implies $\int_0^\infty g_a^2(r, t)r^{n-1}\,dr = \int_0^1(1 - \exp(-r^2 t))^2 r^{-2a-4}r^{n-1}\,dr$ for the radial function g_a.
Answer: $a \in (n/2 - 2, n/2)$.
AMM 85, 1978, 600–601, 6055, S. Zaidman; J. A. Boa.

8.90   Hint: compute the Fourier transform of $e^{-a|x|}$.
For example [R-S] XIII, 138; see [E] 10.9.

8.91 **Answer:** of course the uniqueness does not hold: for $f(x) = \varepsilon_n \sin 2\pi x$, $n \leqslant x < n + 1$, such that $\Sigma \varepsilon_n = 0$, $\Sigma |\varepsilon_n| < \infty$, we have $f(n) = \tilde{f}(n) = 0$. *AMM 89*, 1982, 342–343, 6319, C. Ryavec; I. I. Kotlarski.

8.92 **Hint:** use the Poisson summation formula from the preceding problem. Another method: consider the auxiliary function $f(x) = \Sigma_{k \in \mathbb{Z}} \exp(-(x - k)^2/2t)$, $t > 0$.
H. Dym, H. P. McKean, *Fourier Series and Integrals*, Academic Press, New York, 1972, 1.7.5.
See some generalizations in *AMM 93*, 1986, 822–823, 6491, J. Borwein; W. A. Newcomb, W. Schempp.

8.93 For example [S] I, 50.2; *AMM 87*, 1980, 193–197, A. Steiner; *AMM 79*, 1972, 495–499, M. Pollard, O. Shisha; *Tôhoku Math. J. 24*, 1972, 121–125, R. P. Boas.

8.94 **Hint:** use the decomposition of $1/P(x)$ into the sum of simple fractions.
Answer: $2m - 2$.
[K-G] 682.

8.95 *AMM 68*, 1961, 182, 4890, H. S. Shapiro.

8.97 **Hint:** $\widehat{f''}(y) = -y^2 \hat{f}(y)$.
[H-S] 21.61.

8.98 **Hint:** compute $\int_0^\pi F(t) \sin t \, dt$.
AMM 74, 1967, 330, 5367, D. J. Newman.

8.99 [R-S] XI.149; [E] 12.44.

8.100 [G] 116.

8.101 **Hint:** let $D_K(t) = (2K + 1)^{-1} \Sigma_{k=-K}^{K} e^{ikt}$ be the normalized Dirichlet kernel. Then $f(K, N) = \iint_{\mathbb{T}^2} D_K(x)D_N(y) \, dm(x, y)$ and both iterated limits exist and are equal to $m((0, 0))$.
Answer: to the second part: no.
[R3] 8.14, 8.15.

8.102 **Hint:** $\lim_{n \to \infty} N^{-1} \Sigma_{n=0}^{N-1} |(x, U^n y)|^2 = \Sigma_j |(x, P_j y)|^2$ for all vectors x, y.
[R-S] Th.XI.114; XI.148.

8.103 [R3] 11.17.

8.104 **Hint:** apply to the integral $\int_{\mathbb{R}} |f|^2 = \int_{-\infty}^0 |f|^2 + \int_0^\infty |f|^2$ the inequalities of Hölder, Hardy and Hausdorff-Young consecutively.
For example [Ge] 132 ($p = 2$); for supplementary results see *Int. J. Math. and Math. Sci. 9*, 1986, 185–192, H. P. Heinig, M. Smith.

Answers to Chapter 9
Functional Analysis

9.1 **Hint:** the Cramer formula would be useful.

9.2 **Remark:** see also 9.55, 9.56.

9.3 **Hint:** one may use differential forms.
Answer: $\operatorname{Tr} A \cdot \det(c_1, \ldots, c_n)$.

9.4 **Hint:** observe that this is the Gram determinant of the system of linearly independent functions (x^m), $m = 0, \ldots, n$, considered as the elements of $L^2(0, 1)$.

9.5 *AMM 94*, 1987, 878–879, E3069, Zhang Zaiming; P. Y. Wu, see [P-S] II.68.

9.6 **Answer:** $\Pi_{1 \leqslant p < q \leqslant r} (x_q - x_p)^{n_p n_q}$.
AMM 91, 1984, 586–587, E2942, V. Norton; see *Arch. Math. 34*, 1980, 428–439, Th. Grundhöfer, H. Lüneburg.

9.8 **Answer:** $2^{n(n-1)} \Pi_{i=1}^n \sin^2 t_i \, \Pi_{i<j} (\sin^2 t_i - \sin^2 t_j)$.
AMM 84, 1977, 574, E2589, J. Sunday; J. B. Wilker.

9.9 [Be] 4.25; see also 13, Section 8.3, M. Parodi; [La] 2.11.

9.10 [Be] 6.52, W. Rudin.

9.11 **Hint:** consider the functions $\varphi(z) = \Pi_{k=1}^{n} (z - a_k)/(1 - \overline{a_k}z)$ and the spaces H^2, φH^2.
AMM 84, 1977, 744–745, 6072, W. Lawton; F. Holland.

9.12 **Answer:** for $n > 3$ this implication is not true in general. However if $a_{ij} \leqslant 0$ for $i \neq j$, then the answer is positive. See [Be] 4.37.

9.13 *NAW 20*, 1972, 80-81, 281, A. J. Bosch; A. A. Jageers, D. Kijne.

9.14 **Hint:** recall a condition on eigenvalues of the matrix guaranteeing its diagonalizability.
Remark: as far as we know, it is an open problem: whether for any $\varepsilon > 0$ there exists $C = C(\varepsilon)$ such that for every invertible matrix A, $\|A\| < 1$, there exist an invertible matrix S and a normal matrix N with $\|S^{-1}AS - N\| < \varepsilon$ and $\|S^{-1}\| \, \|S\| \leqslant C$; C is to be independent of n.

9.17 [Be] 6.8.

9.18 [CHM] 3/1961, A.4.

9.19 **Hint:** construct a basis in L using the cyclic vector of A. The coefficients a_j may be determined from the characteristic polynomial of A.
[K-G] 715, 719.

9.20 **Hint:** proof by induction.
[R-N] VI.103.

9.22 For example T. Kato, *Perturbation Theory for Linear Operators*, Springer, 1966, 38–43; [R-S] XII.4.

9.23 **Hint:** consider $\int_0^t Y(s)\, ds$.
Answer: $Y(t) = \exp(tA)$. For example [Be] 10, Section 16.

9.24 **Hint:** $d \exp(tA)/dt = A \exp(tA)$. [Be] 11.63.

9.25 **Hint:** $A = (A + cI) - cI$, $c > 0$. [Be] 10, Section 15.

9.26 [Be] 8.12.

9.27 [Be] 3, Section 7.3, 4.

9.28 **Hint:** an idea of one of numerous proofs: Show that $\det H \leqslant \Pi_{j=1}^{n} h_{jj}$ for every Hermitean positive matrix $H = (h_{ij})$. Use the Jensen inequality and the concavity of the functions $\log \det H$, $(\det H)^{1/n}$. This method is presented in *AMM 89*, 1982, 687–688, A. W. Marshall, I. Olkin. See the next problem with $p = 0$.

9.29 *AMM 92*, 1985, 148–149, J. Borwein.

9.30 [R-S] XIII.169.

9.31 **Hint:** one may use the complex interpolation theory based on the Hadamard three lines theorem.
For example [R-S] IX.4.1, IX.33, Appendix to IX.4.

9.32 **Answer:** no, consider for example the matrices

$$A = \begin{bmatrix} 1 & t \\ t & t^2 \end{bmatrix}, \qquad B = \begin{bmatrix} -1 & t \\ t & -t^2 \end{bmatrix}.$$

9.33 **Hint:** there exists a matrix Q such that $A = QQ^T$, $B = QDQ^T$ for a diagonal matrix D.

9.34 **Hint:** use the Hölder inequality and the formula $\int_0^\infty \cdots \int_0^\infty e^{-(Cx,x)} dx_1 \cdots dx_n = (\pi^n/\det C)^{1/2}$.

9.39 [Be] 14.7; [G-L] VII.127.

9.36 This result is due to Lappo-Danilevski, see for example [Be] 12.8.

9.37 [CHM] 4/1950, II, A.3 or any textbook on finite Markov chains.

9.38 *NAW 24*, 1976, 205, 418, M. L. J. Hautus; J. J. A. M. Brands.

9.39 [Be] 10.1.

9.40 **Hint:** consider $|y(t)|^2$.

9.41 *AMM 91*, 1984, 573–575, R. Koch.

9.42 *AMM 89*, 1982, 586–587, G. A. Fredricks.

9.43 **Answer:** yes.

9.44 **Answer:** for $z = 0$ $A_1 = (\|x\|^2 \|y\|^2 - |(x, y)|^2)^{1/2}/2$, $A_2 = (\|x\|^2 \|y\|^2 - \text{Re}(x, y))^{1/2}/2$.
AMM 81, 1974, 786–787, 5913, D. E. Daykin, J. K. Dugdale; J. R. Kuttler.

9.45 **Hint:** $\Sigma_{n=1}^\infty e_{-n}/n \in L^2(-\pi, \pi)\backslash(X_1 + X_2)$. For example [R3] 1.20; 5.9 or an analogous example in [Di] 6.5.2. See 9.136.

9.46 **Hint:** Let $X_1 = X_3 + \text{lin}\{e\}$, where $e \in \overline{\text{lin}(X_2 \cup X_3)}\backslash(X_2 + X_3)$ and $X_1 \cap X_2 = \{0\}$. This example shows that the lattice of all closed subspaces of any infinite dimensional Hilbert space is not a modular lattice.

9.48 *NAW 14*, 1966, 159, 93, A. C. Zaanen; L. G. Barendregt.

9.49 **Answer:** no. [Ba] 2.5.1.

9.50 For example [G] 349.

9.51 **Answer:** yes, for $p = \infty$ (see the preceding problem), no, for $2 < p < \infty$. Consider the subspace of lacunary series for the construction of a counterexample.
AMM 72, 1965, 435–436, 5199, H. S. Shapiro; J. Williams.

9.52 [K13] 3.40; *Proc. AMS 13*, 1962, 312–314, G. Klambauer.

9.53 **Hint:** show that $v(x)\exp(-x^2) = 0$ a.e. or use the functions $(x \pm i)^{-n} \in C_0(\mathbb{R})$ and the formula $(x + i)^{-1} = i\int_0^\infty \exp(-s + isx)\, ds$. The Hermite polynomials satisfy the identity
$H_n(x)\exp(-x^2/2) = (-1)^n (2^n n!)^{-1/2} \exp(x^2/2)(d/dx)^n(\exp(-x^2))$.
For example [R-S] IX.6.

9.54 **Hint:** consider the subspace $X = (\mathrm{lin}\{g\})^\perp$, where $g \in L^2$ but g is not equivalent to any R^2 function. Observe that $R^2 \cap X$ is dense in R^2. Take an orthonormal system in $R^2 \cap X$. The condition (v) is satisfied but (ii) is not valid.
AMM 65, 1958, 343–345, J. M. H. Olmsted.

9.55 For example [R-N] V.86 or *AMM 75*, 1968, 750, N. Tsao.

9.57 **Hint:** use the Paley-Wiener theorem. [R-N] V.86.

9.59 [AKR] X.33; see problem 3.125.

9.60 **Hint:** observe that (f_n) is complete if and only if for every $x\ \Sigma_{n=1}^\infty \left(\int_0^x f_n(t)\, dt\right)^2 = x$.
AMM 82, 1975, 770, 5957, D. Marsal; P. Chauveheid.

9.61 [G] 164, 336; see [R2] 4.10.

9.62 For example [G] 337.

9.63 [R-S] V.49.

9.64 **Answer:** $s\text{-}\lim_{n\to\infty}(PQ)^n$ = the orthogonal projection onto $X \cap Y$. For example [R-S] VI.39; [K-G] 758.

9.67 **Hint:** consider the shift on ℓ^2, $S(x_1, x_2, \ldots) = (0, x_1, x_2, \ldots)$.
Answer: if and only if the space is finite dimensional. For example [AKR] X.73.

9.68 **Answer:** $\sigma(S) = \{0\}$, $\|S^n\| = \exp(-(n-1)n(n+1)/12)$. [R3] 13.18.

9.69 For example [Do] 5.6, 5.8.

9.70 [G] 347; see 9.125.

9.71 **Hint:** use the standard M. Riesz compactness criterion ([R-S] Th. XIII.65, XIII.117, Th. XIII.66): The closure of the subset $C \subseteq L^2(\mathbb{R}^n)$ is compact if and only if the following conditions are satisfied:

(i) For every $\varepsilon > 0$ there exists a compact set $K \subseteq \mathbb{R}^n$ such that $\int_{\mathbb{R}^n \setminus K} |f|^2 \leqslant \varepsilon$ for all $f \in C$.

(ii) For every $\varepsilon > 0$ there exists $\delta > 0$ such that $\int |f(x - y) - f(x)|^2 \, dx \leqslant \varepsilon$ for all $|y| \leqslant \delta$ and $f \in C$.

9.72 **Hint:** how can one characterize this eigenvalue knowing the eigenvector expansion of x?
[Di] 11.5.10.

9.73 For example [G] 330; [S-K] 8.3.13.

9.74 [R3] 4.14, 10.19.

9.75 **Hint:** assume that such S exists. Then $x_1 = Se_1 = 0$, $x_2 = Se_2 = ce_2$ for some scalar c and $x_2 = 0$, which contradicts $Te_2 \neq 0$. To prove this observe that $\overline{T^*(H)} \subseteq \overline{S^*(H)} \perp x_1$. [Di] 11.5.13.

9.76 **Hint:** $\|T\|_{2,2} = 2$ (and similarly $\|T\|_{p,p} = p/(p - 1)$) which follows from the Hardy inequality, see for example [R2] 3.14. [R3] 4.17; [K-G] 388.

9.77 **Answer:** $\|T\| = 2$. *NAW 20*, 1972, 89–90, 288, P. van der Steen.

9.78 For example [Do] 5.28.

9.79 **Hint:** each operator A can be approximated by the operators with matrices

$$
\begin{bmatrix}
A_n & I_n & 0 & & \\
I_n - A_n^2 & -A_n & & & \\
 & & 1 & & \\
0 & & 1 & & \\
 & & & & \ddots
\end{bmatrix}
$$

where A_n corresponds to $n \times n$ part of the matrix of A.
AMM 82, 1975, 309–310, 5944, J. L. Wallen.

9.80 **Answer:** (i) if $f(g) = \overline{f(-g)}$, (ii) if $|\hat{f}(x)| = 1$ which is possible only for the discrete group, (iii) if G is a compact group.
[K-G] 793.

9.81 **Hint:** consider K in the Fourier representation. Recall the properties of the Fourier transform of $k \in L^1$ with compact support.
Answer: (i) K is never invertible, (ii) only for $k \equiv 0$. For example [H-P] 3.16, 3.17.

9.83 **Hint:** let the subset $I \subseteq \mathbb{N}$ have n elements. Define A on the subspace V spanned by e_j, $j \in I$, so that $(Ae_i, e_j) = n^{-1/2}$ for i, $j \in I$. Observe that $\|A_{|V}\| \geq n^{1/2}$.
[Di] 6.5.4b.

9.84 **Hint:** it suffices to consider the case $a_{ij} = 0$ if $j - i \neq p$, $p \in \mathbb{N}$ fixed.
[Ba] 3.4.1.

9.85 **Answer:** no.
AMM 69, 1962, 814–815, 4971, D. J. Newman, N. G. de Bruijn.

9.86 **Hint:** one of the methods of proof consists in using the Nehari lemma:
If $c_k \in \mathbb{C}$, $m_k \in \mathbb{Z}$, $x \in \mathbb{R}$, $x \neq m_k$, then $\int_0^{2\pi} |\Sigma_{k=1}^n c_k \exp(im_k t)| \, dt \geq 2|\sin \pi x|$ $|\Sigma_{k=1}^n c_k/(x - m_k)|$. For its proof observe that $|\Sigma_{k=1}^n c_k \exp(im_k t)| = |\Sigma_{k=1}^n c_k \exp(i(m_k - x)t)|$.
Then apply this lemma with $x = 1/2$ and $(\Sigma_{k=1}^n |a_k| \exp(ikt))(\Sigma_{k=1}^n |b_k| \exp(ikt))$ instead of $\Sigma_{k=1}^n c_k \exp(im_k t)$.
[He] IV.5.4; see also [Be] 5.10 and other interesting properties of the Hilbert matrix in *AMM 90*, 1983, 301–312, Man-Duen Choi.

9.87 **Hint:** if b_{ij} are the entries of $A^T A$, then $b_{ij} < 1/(i + j)$ hence $A^T A$ is bounded.
AMM 69, 1062, 240, 4947, D. J. Newman.

9.88 [Hi2] 10.5.6.

9.89 **Answer:** one of these ranges is equal to H^2, the other is of codimension 1. In the second case the defect indexes of T are equal to 0 and ∞.
[R3] 13.9.

9.90 **Answer:** the index is equal to the sum of multiplicities of all zeros of g in G.
[K-G] 403.

9.91 **Answer:** For example $A(x_1, x_2, x_3, \ldots) = (x_1, x_2^2, x_3^3, \ldots)$ is not bounded on the balls of radii greater than 1.

9.92 **Answer:** the classical example of S. Kakutani is $A(x_1, x_2, x_3, \ldots) = ((1 - \|x\|^2)^{1/2}, x_1, x_2, \ldots)$.

9.93 **Hint:** first show the J. P. Vigier lemma: Every monotone and bounded sequence of symmetric operators strongly converges to some symmetric operator. Then consider $B = I - A$, $Y_n = I - X_n$.
[R-N] VII.104; [AKR] X.115; *NAW 15*, 1967, 173–174, 138, A. C. Zaanen; P. van der Steen.

9.94 **Hint:** without loss of generality one may assume that $0 \leqslant A \leqslant I$. The recurrent sequence $X_0 = 0$, $X_{n+1} = X_n + k^{-1}(A - X_n^k)$ converges to the kth root of A.
For example *AMM 89*, 1982, 313–314, N. M. Rice; *AMM 87*, 1980, 380–382, D. R. Brown, M. J. O'Malley.

9.95 For example [R-S] VI.30.

9.96 [W] 4.4. Example 7. The construction is due to P. D. Lax.

9.97 **Answer:** let (e_i) be an orthonormal basis. After rearranging (e_i) in a double sequence (e_{ij}) define $Te_{ij} = e_i$ for all j.
AMM 68, 1961, 938, 4926, S. Berberian.

9.98 **Answer:** for example $Te_n = ne_1$, where (e_n) is an orthonormal basis. The couple $(0, e_1)$ belongs to the closure of the graph of T.
   Another example: consider the adjoint operator for $Tf = (\int_{\mathbb{R}} f)g$, where $f \in L^1(\mathbb{R}) \cap L^2(\mathbb{R})$, $g \in L^2(\mathbb{R})$, $g \neq 0$.
See [R-S] VIII.1; [S-K] 10.3.4.

9.99 **Hint:** $(I + A^*A)^{-1}$ is a bounded operator.
[K-G] 775.

9.101 [R-S] VIII.5.

9.102 *AMM 93*, 1986, 659–660, 6487, Pei Yuan Wu; Z. Franco.

9.103 *AMM 94*, 1987, 197–199, 6496, R. Szwarc.

9.104 **Hint:** observe that P and Q commute with $(P - Q)^2$ and therefore with the spectral measure of $(P - Q)$, see A. Connes, *The Chern character in K-homology* (preprint); a simple proof was given by U. Haagerup (unpublished).

9.105 [R-N] VIII.125.

9.106 **Answer:** yes, in the case of shift. [AKR] X.161; [K-G] 757, [S-K] 9.2.1.

9.107 **Hint:** let $f_n(x) = 1/x$ for $x \geqslant 1/n$ and n for $x \leqslant 1/n$. Then $U = s\text{-}\lim_{n \to \infty} Af_n(|A|)$.
[R-S] VII.20.

9.109 **Answer:** no, for example let A be defined in $L^2(\mathbb{R})$ as $Af(x) = f(x) + (1 + x)f(-x)$ if $x \geqslant 0$ and $-f(x)$ if $x < 0$, $D(A) = \{f : \int_0^\infty |f(x)|^2 \, dx + \int_{-\infty}^0 (1 + |x|^2)|f(x)|^2 \, dx < \infty\} \subseteq L^2(\mathbb{R})$. Observe that $0 \in \sigma(A)$.
AMM 72, 1965, 92, 5166, P. Rejto, C. Conley; F. P. Callahan.

9.110   [R-S] XII.19.

9.111   **Hint:** for every x there exists a finite rank operator F of small norm such that $H + F$ has a finite dimensional reducing subspace which almost contains x. For example, [Do] 5.17.

9.112   For example [Do] 5.18.

9.113   **Answer:** ind $A_{\pm} = \pm 1$.
[K-G] 407.

9.114   *NAW 21*, 1973, 291, 339, R. A. Hirschfeld; M. L. J. Hautus.

9.116   **Hint:** consider the function $[0, 1] \ni t \mapsto \exp(i(tA + (1 - t)B))$.

9.117   For example [G] 389.

9.118   **Hint:** assume contrarily $X^* = \bigcup_n \{x^*: x^*(x_n) = 0\}$.

9.119   **Answer:** $A = (a_{ij}) = (2^{-i}\delta_{ij})$, $B = (b_{ij})$ where $b_{0j} = 2^{-j}$, $b_{ij} = b_{ji} = \delta_{ij}$ for $i \geqslant 1$. The equation $ABx = y$ has no solution for $y = (1)$.
AMM 71, 1964, 445–446, 5094, A. Wilansky; S. T. Ackermans.

9.120   [K-G] 333.

9.121   **Answer:** yes, for example for $f(t) = 1/(1 + t)$. For the second question see problem 7.138.

9.122   **Answer:** $h(f) = \int_0^1 f(t)\, dt$. The uniqueness follows from the convergence of $h(S^n f)$ to $\int_0^1 f$.

9.123   **Hint:** determine the extreme points of the unit ball in $C[0, 1]$. Compare this to the Banach-Alaoglu and Krein-Milman theorems. For example [K-G] 299, 378, 437.

9.125   [G] 75; see 9.75.

9.126   *AMM 74*, 1967, 772–773, Putnam Competition 1966, A-5.

9.127   **Hint:** the norms from C^1 and C are equivalent on X. For example [G] 90.

9.128   **Hint:** consider piecewise linear functions with values equal to 1 at q_n, (q_n) the sequence containing all the rationals in $[0, 1]$. For example [K-G] 459.

9.130   *AMM 87*, 1980, 39–40, A. M. Russell; see [H-S] Section 18.

9.131   **Answer:**   for   example $M = I \times I$,   where $I = [0, 1]$, $v(E) = \Sigma_{x \in I} m(\{y \in I : (x, y) \in E\})$, $\mu(E) = v(E) + \Sigma_{y \in I} m(\{x \in I : (x, y) \in E\})$ for   all   Borel subsets $E \subseteq M$. Define the functional on $L^1(M, \mu)$ by $\int_M f\, dv$.

Its norm is less than 1 but it cannot be represented as $\int fg\,d\mu$ for any bounded function g.
[H-S] 20.17.

9.132 Answer: if and only if $p = 2n$, $n \in \mathbb{N}$. *AMM 89*, 1982, 592, S 33, S. Reich; B. Reznick.

9.133 Answer: yes, similarly as in each strictly convex Banach space. This implication is not true in ℓ^1, ℓ^∞: take $x = (1/2, 1, 0, \ldots)$, $y = (0, 1/2, 0, \ldots)$.

9.134 For example [R-S] III.25; [H-S] 15.5, 15.8.

9.135 Hint: use the Clarkson inequality.
Remark: for $p = 1$ this result is not true: $f_n = 1 + \sin nx \to 1$ weakly in $L^1(0, 2\pi)$, but f_n does not converge neither in measure nor in L^1. For example [H-S] 15.17, 15.18; [R-N] II.37.

9.136 Answer: no.

9.137 [R3] 9.15.

9.139 Answer: (i) $\exp(ix)$, (iii) $\exp(i \log x)$ is an eigenfunction of T.
AMM 73, 1966, 790–791, 5315, G. A. Heuer, R. G. Lee; J. T. Rosenbaum, R. O. Davies.

9.140 Hint: consider (i) piecewise linear functions, (ii) step functions.
[K-G] 429.

9.141 [R-S] IV.12.

9.142 [R-S] III.32.

9.143 Hint: the set $\{f : \int |f|^2 \leqslant n\}$ (closed in L^1) has no interior points,
or: $\int fg_n \to 0$ for $f \in L^2$ if $g_n(t) = n$ on $[0, n^{-3}]$, $g_n(t) = 0$ otherwise, but this convergence does not hold for some L^1 functions,
or: the imbedding $L^2 \subseteq L^1$ is continuous but not surjective.
[R3] 2.4; the original paper: *Bull. AMS 43*, 1937, 245–248, J. C. Oxtoby.

9.144 Hint: consider the zero functional defined on the subspace of ℓ^∞ containing all the sequences of the form $(x_0, x_1 - x_0, x_2 - x_1, \ldots)$. Find an extension L of this functional such that $L(1, 1, \ldots) = 1$, $\|L\| = 1$.
For example *AMM 74*, 1967, 308–311, L. Sucheston.

9.145 Hint: h is positive and hence, continuous. It suffices to verify the multiplicative property on the dense subalgebra of simple functions in ℓ^∞. Another method: observe that if x, $1/x \in \ell^\infty$, then $h(x) \neq 0$ and apply the Gleason-Kahane-Żelazko theorem (for example [R3] Th. 10.9).
AMM 86, 1979, 704–705, E2712, A. Wilansky; P. R. Chernoff, J. W. Lark III.

9.146   **Hint:** one may use the Dunford-Pettis theorem: Each operator T from a separable Banach space Z into $L^\infty(\mathbb{R})$ is of the form $Tz = \int g(x)(z)\,dx$, where $g: \mathbb{R} \to Z^*$ is a measurable function, $\sup_{x\in\mathbb{R}} \|g(x)\| = \|T\|$.
Examples: for $p \geqslant 2$ consider the bilinear form $\int G(x)f(x)g(x)\,dx$, where $G \in L^r$, $1/r + 2/p = 1$. For $1 < p \leqslant 2$ consider $\int G(x - y)f(x)g(y)\,dxdy$, where $G \in L^r$, $1/r = 2(1 - 1/p)$; use the Young inequality $\|k * g\|_s \leqslant \|k\|_r \|g\|_p$ with $1 + 1/s = 1/p + 1/r$.
[R-S] V. 33.

9.147   **Hint:** the limit g of a net of translates of f satisfies the functional equation $g(s + t) = g(s)g(t)$ but g is not necessarily continuous.
AMM 90, 1983, 394–396, L. A. Rubel, A. Siskakis.

9.148   **Hint:** K is the intersection of all nonempty compacts C with the property that there exists $\delta > 0$ such that for all $f \in C(X) \, |f(x)| < \delta$ and $x \in C$ imply $|h(f)| < 1$.
AMM 85, 1978, 56–57, 6093, R. Johnsonbaugh; F. Michael Christ.

9.149   For example *AMM 82,* 1975, 861, 5961, S. Zaidman, S. J. Sidney.
Remark: the property stated in this problem is called the accretivity of $-A$.

9.150   [Hi2] 7.1.4; [Ox] Ch. 11.

9.151   [Hi2] 5.3.8–12.

9.152   **Hint:** consider the following semigroups $T(t)$ in $C[0, \infty]$
   (i) $T(t)f(s) = f(t + s)$.
   (ii) $T(t)f(s) = (f(t + s) + f(t - s))/2$.
[Hi2] 5.3.7, Th. 5.31; see also *AMM 93,* 1986, 444–451, R. Farwig, D. Zwick.

9.154   **Answer:** no, consider in ℓ^2 $B = \{(x_n, y_n) : n \in \mathbb{N}\}$ such that $x_n = (0, 0, \ldots, 0, 1, 0, \ldots)$ (1 on $(n + 1)$th place), $y_n = (1, 0, \ldots, 0, 1, 0, \ldots)$. B is a dual generalized basis not being a generalized basis: $(1, 0, 0, \ldots) \perp x_n$ for every n.
AMM 80, 1973, 325–326, 5826, R. Tapia; E. M. Klein.

9.155   *NAW 24,* 1976, 192–194, 395, D. van Dulst.

9.156   For example *AMM 70,* 1963, 574–575, W. A. J. Luxemburg, A. C. Zaanen; A. K. Snyder, A. Wilansky.

9.157   **Answer:** no, for example Ker h contains a sequence (x_n), $\|x_n\| = 1$, without norm convergent subsequences. For each vector $v \notin$ Ker h the set $S = \{x_n + n^{-1}v : n \in \mathbb{N}\}$ is closed and $h(S)$ is obviously not closed.
Remark: evidently, the image of every bounded closed and convex set is compact.
AMM 84, 1977, 829–830, 6078, A. Wilansky, S. Goldberg, R. C. James. See S. Goldberg, *Unbounded Linear Operators,* McGraw-Hill, New York, 1976, IV.1.10.

9.158 **Answer:** if the dual space is strictly convex.
Duke Math. J. 5, 1939, 538–547, A. E. Taylor; *Proc. AMS 9*, 1958, 325, S. R. Foguel; see also *Trans. AMS 95*, 1960, 238–255, R. R. Phelps.

9.159 [R-N] 31.87; [H] 27. See a discussion on this subject in the note *AMM 87*, 1980, 217, J. Hennefeld and *AMM 90*, 1983, 476.

9.160 **Hint:** consider a dense subset (x_n) of the unit sphere in the space and the operator A from ℓ^1 defined by $A((a_n)) = \Sigma_{n=1}^{\infty} a_n x_n$. Identify the space with $\ell^1/\mathrm{Ker}\, A$.
[R-S] III.29.

9.161 [K-F] IV.6; [AKR] VIII.61.

9.162 **Answer:** no, for example $X = c_0$, $e_n = (0,\dots,\; 0,\; 1,\; 0\dots)$ and $T e_n = n^{-1} e_n$.
AMM 87, 1980, 68, A. Wilansky; G. Gripenberg.

9.163 **Answer:** for example $X = C[0,1]$, $Y = L^{\infty}[0,1]$, $Z = L^1[0,1]$, T, S are the natural imbeddings.
See the reference to the preceding problem.

9.164 [R-S] III.27.

9.165 **Hint:** consider $L^1(0, 2\pi)$ and the operator $\Sigma_{n\in\mathbb{Z}} a_n e^{inx} \mapsto \Sigma_{n=-\infty}^{-1} |a_n/n^3| e^{inx}$.
AMM 73, 1966, 680, 5302, P. Garrett; L. J. Wallen, J. L. Ercolano.

9.166 **Answer:** yes, in the reflexive spaces, no in general.
Example: T is defined in $C[-1,1]$ by $Tf(s) = \int_{-1}^{s} f(t)\,dt$. Consider the functions $f_n(t) = 0$ for $-1 \leqslant t \leqslant 0$, $= nt$ for $0 < t \leqslant 1/n$ and $= 1$ for $1/n < t \leqslant 1$. The limit of $Tf_n \notin \mathrm{Ran}\, T$.
[K-F] IV.6.1.

9.167 [R-S] VI.20.

9.169 **Hint:** use the Stone-Weierstrass theorem. The formula $x - ixi - jxj - kxk = 4a1$ for $x = a1 + bi + cj + dk$ may be useful.
[H-S] 7.4.5, J. C. Holladay.

9.170 *AMM 94*, 1987, 195–196, 6494, A. Wilansky; A. A. Jagers.

9.171 **Answer:** $X(t) = (I + tA)^{-1}$.

9.172 **Answer:** not always, but the answer is positive if some C_n is compact.

9.173 **Hint:** one may suppose that $0 \in C$. Apply the Banach fixed point theorem to the operators $T_r = (1 - r)T : C \to C$, $0 < r < 1$. Then show that

$\|x_r - T(x_r)\| \leqslant r \cdot \text{diam } C$ for (the unique) fixed point x_r of T_r. Hence $\inf_{x \in C} \|x - T(x)\| = 0$ and this infimum is attained.
AMM 84, 1977, 363–364, W. G. Dotson, W. R. Mann; *AMM 86*, 1979, 840, P. K. F. Kuhfittig.

9.174 *Proc. AMS 10*, 1959, 448–450, W. Cheney, A. A. Goldstein.

9.175 **Answer:** (i) there exists such a nontrivial function, (ii) f vanishes identically.
[K-G] 684; [K-F] VIII.4.4.

9.176 **Hint:** the differentiation operator is closed (in a suitable function space).

9.177 [S2] Section 2.4, 5, 6.

9.178 **Answer:** for $n = 2$ consider $u(x) = |\log |x| \, |^{1/4}/(1 + |x|)^2$, $x \in \mathbb{R}^2$.
[R3] 7.21, 8.11.

9.179 [R-S] V.16.

9.180 [R-S] IX.59, Ex. 1, IX Section 10.

9.181 **Hint:** the integration by parts is not valid for the integral $\int \exp(ie^x)\varphi(x) \, dx$ for some $\varphi \in \mathscr{S}$.
[K-G] 519.

9.182 **Answer:** for example $L\varphi = \sum_{n=1}^{\infty} D^n \varphi(1/n)$ for $\varphi \in \mathscr{D}(0, \infty)$.

9.183 **Answer:** in the first case $T = \sum_{p_1=0}^{m_1} c_{1p_1} \delta_{a_1}^{(p_1)} + \cdots + \sum_{p_k=0}^{m_k} c_{kp_k} \delta_{a_k}^{(p_k)}$ is the general solution, c_{jp_j} are arbitrary constants. In the second case S/P is a particular solution. For example [CB] II.2.

9.184 **Answer:** $\pi\delta$, $-\pi\delta$, $\pi\delta'$. For example [CB] II.5; [R-S] V.22.

9.185 **Answer:** $(2 \text{ p.v. } x^{-1})'$.
[CB] II.9.

9.186 **Answer:** $u_\varepsilon = (1 + 4\varepsilon)^{-1/2}(Y(x) \exp(r_2 x) + Y(-x) \exp(r_1 x))$, where $r_{1,2} = (1 \pm (1 + 4\varepsilon)^{1/2})/2\varepsilon$, $u_0 = Y(x) \exp(-x)$ and Y is the Heaviside function.
[CB] II.12.

9.187 **Answer:** given arg a there exists at most one v such that the nonzero limit of $f_{a,v}$ exists (and it is a multiple of δ; consider the relation $f'_{a,v} = xf_{a,v}$).
   Given v there exist at most two values of p such that the limit $g_{a,v,p}$ is not 0 ($p = 2v - 1$, $p = 2v - 2$) and this limit is a multiple of δ or δ'. Begin with, for example, the case $v = 3/2$, remember that $g'_{a,v,p} = pg_{a,v,p-1} + 2ag_{a,v,p+1}$.
[CB] II.16.

9.188   Hint: consider the distribution f such that $(f, \varphi) = \int f(x)[\varphi(x) - (\varphi(0) + x\varphi'(0) + \cdots + x^{m-1}\varphi^{(m-1)}(0)/(m-1)!)h(x)]\,dx$, where h is a smooth function equal to 1 for $|x| \leqslant 1$ and 0 for $|x| \geqslant 2$.
[S2] Section 3.4, 5. See a multidimensional analog in Sections 6.6, 7 and Sections 11.1, 15.

9.189   Hint: observe that the classical solution is $C\exp(1/x^2)$. Apply the result of the preceding problem.
[S2] Section 5.2.

9.190   Hint: define first (y, ψ) for test functions ψ of the form $\psi(x) = (x\varphi(x))'$, $\varphi \in \mathscr{D}$, that is $\psi \in \mathscr{D}$ such that $\int_{-\infty}^{\infty} \psi = 0 = \int_{-\infty}^{0} \psi$.
[S2] Section 5.3.

9.191   Hint: study the action of T on the translates of $f * g$, write down a differential equation satisfied by T. Observe that the support of S is a singleton.
Answer: $T = e^{cx}$, $S = \delta_y$ for some $y \in \mathbb{R}$.
[CB] III.6.

9.192   Answer: in $\mathscr{D}'$ all the sequences (a_n), in $\mathscr{E}'$ if $a_n = 0$ for sufficiently large n.
[CB] II.8, IV.12.

9.193   Answer: these series converge in $\mathscr{D}'$ and diverge in $\mathscr{E}'$ for all $a > 0$. The first series converges in $\mathscr{S}'$ for all $a \leqslant 1$, the second only for $a = 1$. The Fourier transforms are equal to $(1 - ae^{iy})^{-1}$ and for $a = 1$ $1 - d^2(\Sigma_{n=1}^{\infty} e^{iny}/n^2)/dy^2$ and $1 - 2d^2(\Sigma_{n=1}^{\infty} (\cos ny)/n^2)/dy^2$ respectively.
[CB] IV.3.

9.194   Answer: T is always defined in $\mathscr{D}'$ and in $\mathscr{S}'$ if and only if $|s_k| \leqslant C(1 + k^p)$ for some $C, p \geqslant 0$. For $T = \delta_0 - 2\delta_1$, $E = \Sigma_{n=0}^{\infty} 2^n \delta_n \notin \mathscr{S}'$.
[CB] IV.12.

9.195   Answer: $|c_n| \leqslant C(1 + |n|^p)$ for some $C, p \geqslant 0$.
[K-G] 525.

9.196   [R3] 9.2.

9.197   [S2] Section 7.9.

9.198   Hint: prove that $|(f, \varphi(x + h))| \leqslant C\|\varphi\|_m(1 + |h|^p)$ for some m, p (independent of φ!). Then represent the function from $\mathscr{S}$ as the series of the functions from $\mathscr{D}$. A function $\psi \in \mathscr{D}$ such that $\Sigma_{n=0}^{\infty} \psi(x + h_n) = 1$ for some $h_n > 0$ would be useful.
[S2] Section 10.8.

9.199   [S2] Section 7.5.

9.200   For the (im)possibility of multiplication of distributions see P. Antosik, J. Mikusiński, R. Sikorski, *Theory of Distributions. The Sequential Approach*, PWN, Warsaw and Elsevier, Amsterdam, 1973.

References

A. Well-known textbooks, monographs and problem books with their contents (nearly) disjoint with this book

[B]   R. P. Boas, *A Primer of Real Functions*, MAA, 1972.

[Bm]   T. J. I'A. Bromwich, *An Introduction to the Theory of Infinite Series*, Macmillan, London, 1965.

[D]   B. P. Demidovič, *Sbornik zadač i upražnenij po matematičeskomu analizu*, Nauka, Moskva, 1969.

[OD]   *The Otto Dunkel Memorial Problem Book*, New York, 1957, supplement to the American Mathematical Monthly 64/7, 1957; ed. H. Eves, E. P. Starke.

[F]   G. M. Fichtengolc, *Kurs differencialnogo i integralnogo isčislenija*, Nauka, Moskva, 1969.

[Fe]   W. Feller, *An Introduction to Probability Theory and its Applications*, John Wiley & Sons, New York, 1966.

[G-O]   B. R. Gelbaum, J. H. M. Olmsted, *Counterexamples in Analysis*, Holden Day, San Francisco, 1964.

[Ge]   C. George, *Exercices et problèmes d'intégration*, Bordas, Paris, 1980. Translation: *Exercises in Integration*, Springer, New York, 1984.

[G-L]   I. M. Glazman, Ju. I. Ljubič, *Konečnomernyj linejnyj analiz*, Nauka, Moskva, 1969. Translations: *Analyse linéaire dans les espaces de dimensions*

218

finies, Mir, Moscou, 1972; *Finite-Dimensional Analysis*, MIT, Cambridge, 1974.

[H]   P. R. Halmos, *A Hilbert Space Problem Book*, D. Van Nostrand, Princeton, 1967 and (second edition) Springer, New York, 1982.

[Kl1]   G. Klambauer, *Problems and Propositions in Analysis*, Marcel Dekker, Inc. New York, 1979.

[Kn]   K. Knopp, *Theorie und Anwendungen der Unendlichen Reihen*, Springer, Berlin, 1947. Translation: *Theory and Application of Infinite Series*, Blackie and Son, London, 1966.

[K]   J. Krzyż, *Zbiór zadań z funkcji analitycznych*, PWN, Warszawa, 1972. Translation: *Problems in Complex Variable Theory*, American Elsevier, New York and PWN, Warszawa, 1971.

[P-S]   G. Pólya, G. Szegö, *Aufgaben und Lehrsätze aus der Analysis*, Springer, Berlin, 1964. Translation: *Problems and Theorems in Analysis*, Springer, Berlin, 1972, 1976.

[R-N]   F. Riesz, B. Sz.-Nagy, *Leçons d'analyse fonctionnelle*, Akadémiai Kiadó, Budapest, 1972. Translation: F. Ungar, New York, 1955.

[R1]   W. Rudin, *Principles of Mathematical Analysis*, McGraw-Hill, New York, 1976.

[R2]   W. Rudin, *Real and Complex Analysis*, McGraw-Hill, New York, 1974.

[S-P]   V. A. Sadovničij, A. S. Podkolzin, *Zadači studenčeskich olimpiad po matematike*, Nauka, Moskva, 1978.

[S-Z]   S. Saks, A. Zygmund, *Analytic Functions*, PWN, Warszawa and Elsevier, Amsterdam, 1971.

[S]   B. V. Šabat, *Vvedenie v kompleksnyj analiz*, Nauka, Moskva, 1985.

[Si]   W. Sierpiński, *Działania nieskończone*, Czytelnik, Warszawa, 1948.

[S1]   G. E. Šilov, *Matematičeskij analiz. Funkcii neskolkich veščestvennych peremennych*, Nauka, Moskva, 1972. Translation: *Analyse mathématique. Fonctions de plusieurs variables réelles*, Mir, Moscou, 1975.

[T]   E. C. Titchmarsh, *The Theory of Functions*, Oxford University Press, 1939.

[Z]   A. Zygmund, *Trigonometric Series*, Cambridge University Press, 1968.

B.   Effectively cited sources

Journals

AMM	*American Mathematical Monthly*
CMB	*Canadian Mathematical Bulletin*
D	*Delta* (in Polish)
EM	*Elemente der Mathematik*
K	*Kvant* (in Russian)
MS	*Matematika v škole* (in Russian)
MM	*Mathematical Magazine*
NAW	*Nieuw Archief voor Wiskunde*
RMS	*Revue des Mathématiques Spéciales*

Books

[A] G. Alexits, *Convergence Problems of Orthogonal Series*, Akadémiai Kiadó, Budapest and Pergamon Press, Oxford, 1961.

[AKR] A. B. Antonevič, P. N. Kniazev, Ja. V. Radyno, *Zadači i upražnenia po funkcionalnomu analizu*, Vyšejšaja škola, Minsk, 1978.

[As] R. B. Ash, *Complex Variables*, Academic Press, New York, 1971.

[Ba] A. V. Balakrishnan, *Applied Functional Analysis*, Springer, New York, 1976.

[Br] N. K. Bari, *Trigonometričeskie riady*, GIFML, Moskva, 1961. Translation: *A Treatise on Trigonometric Series I, II*, Pergamon Press, Oxford, 1964.

[Be] R. Bellman, *Introduction to Matrix Analysis*, McGraw-Hill, New York, 1960.

[B-L] Th. Bröcker, L. Lander, *Differentiable Germs and Catastrophes*, Cambridge University Press, 1975.

[C] H. Cartan, *Calcul différentiel*, Hermann, Paris, 1967. Translation: *Differential Calculus*, Hermann, Paris, 1971.

[Ch] K. Chandrasekharan, *Introduction to Analytic Number Theory*, Springer, Berlin, 1968.

[CB] Y. Choquet-Bruhat, *Distributions. Théorie et problèmes*, Masson, Paris, 1973.

[CHM] *Contests in Higher Mathematics, Hungary 1949–1961*, ed. G. Szász, L. Gehér, I. Kovács, L. Pintér, Akadémiai Kiadó, Budapest, 1968.

[Co] J. B. Conway, *Functions of One Complex Variable*, Springer, Berlin, 1973.

[Dn] P. Dienes, *The Taylor Series*, Dover Publications, New York, 1957.

[Di] J. Dieudonné, *Foundations of Modern Analysis*, Academic Press, New York, 1960.

[Do] R. Douglas, *Banach Algebra Techniques in Operator Theory*, Academic Press, New York, 1972.

[E] R. E. Edwards, *Fourier Series. A Modern Introduction 1–2*, Springer, New York, 1979–1982.

[Ev] M. A. Evgrafov, *Asimptotičeskie ocenki i celye funkcii*, Nauka, Moskva, 1979. Translation: *Asymptotic Estimates and Entire Functions*, Gordon and Breach, New York, 1961.

[G] B. Gelbaum, *Problems in Analysis*, Springer, New York, 1982.

[H-R] G. Hardy, W. W. Rogosinski, *Fourier Series*, Cambridge University Press, 1965.

[He] M. Heins, *Complex Functions Theory*, Academic Press, New York, 1968.

[H-S] E. Hewitt, K. Stromberg, *Real and Abstract Analysis*, Springer, Berlin, 1965.

[Hi1] E. Hille, *Analytic Function Theory I, II*, Ginn and Company, Boston, 1962–1963.

[Hi2] E. Hille, *Methods in Classical and Functional Analysis*, Addison-Wesley, Reading, 1972.

[Hr] M. W. Hirsch, *Differential Topology*, Springer, New York, 1976.

[Ho] K. Hoffman, *Banach Spaces of Analytic Functions*, Prentice-Hall, Englewood Cliffs, New Jersey, 1965.

[H-P] V. C. L. Hutson, J. S. Pym, *Applications of Functional Analysis and Operator Theory*, Academic Press, New York, 1980.

[Ka] J. P. Kahane, *Séries de Fourier absolument convergentes*, Springer, Berlin, 1970.

[K-G] A. A. Kirillov, A. D. Gvišiani, *Teoremy i zadači funkcionalnogo analiza*, Nauka, Moskva, 1979 and 1988. Translations: *Theorems and Problems in Functional Analysis*, Springer, NY, 1982; *Théorèmes et problèmes d'analyse fonctionnelle*, Mir, Moscou, 1982.

[Kl2] G. Klambauer, *Mathematical Analysis*, Marcel Dekker, Inc., New York, 1975.

[Kl3] G. Klambauer, *Real Analysis*, American Elsevier, New York, 1973.

[K-F] A. N. Kolmogorov, S. V. Fomin, *Elementy teorii funkcij i funkcionalnogo analiza*, Nauka, Moskva, 1981. Translation: *Eléments de la théorie des fonctions et de l'analyse fonctionnelle*, Mir, Moscou, 1974.

[K-N] L. Kuipers, H. Niederreiter, *Uniform Distribution of Sequences*, John Wiley & Sons, New York, 1974.

[La] P. Lancaster, *Theory of Matrices*, Academic Press, New York, 1969.

[L] G. Lefort, *Algèbre et analyse. Exercices*, Dunod, Paris, 1968. Translation: *Algebra and Analysis. Problems and Solutions*, North Holland, Amsterdam, 1966.

[Le] A. F. Leontiev, *Celye funkcii. Riady eksponent*, Nauka, Moskva, 1983.

[M] A. I. Markuševič, *Izbrannye glavy teorii analitičeskich funkcij*, Nauka, Moskva, 1976.

[MS] *Matematika segodnia*, ed. A. Ja. Dorogovcev, Višča škola, Kiev, 1983.

[MV] *Mathematikai Versenytételek*, ed. J. Kürschak et al., Tankönyvkiado, Budapest, 1965. Russian translation: Mir, Moskva, 1976.

[N1] R. Narasimhan, *Analysis on Real and Complex Manifolds*, Masson, Paris and North Holland, Amsterdam, 1968.

[N2] R. Narasimhan, *Complex Analysis in One Variable*, Birkhäuser, Boston, 1985.

[New] D. J. Newman, *A Problem Seminar*, Springer, New York, 1982.

[O] Ju. S. Očan, *Sbornik zadač po matematičeskomu analizu*, Prosveščenie, Moskva, 1981.

[Ox] J. C. Oxtoby, *Measure and Category*, Springer, New York, 1980.

[P] G. Pólya, *Mathematics and Plausible Reasoning*, Princeton, 1954.

[Po] M. M. Postnikov, *Vvedenie v teoriju algebraičeskich čisel*, Nauka, Moskva, 1982.

[Pr] I. I. Privalov, *Vvedenie v teoriju funkcij kompleksnogo peremennogo*, Nauka, Moskva, 1977.

[R-S] M. Reed, B. Simon, *Methods of Modern Mathematical Physics*, *1–4*, Academic Press, New York, 1972–1979.

[R3] W. Rudin, *Functional Analysis*, McGraw-Hill, New York, 1973.

[R4]   W. Rudin, *Function Theory in the Unit Ball of* $\mathbb{C}^n$. Springer, Berlin, 1980.

[S-K]   I. E. Segal, R. A. Kunze, *Integrals and Operators*, McGraw-Hill, New York, 1968.

[Sr]   R. Sikorski, *Funkcje rzeczywiste I, II*, PWN, Warszawa, 1958–1959.

[S2]   G. E. Šilov, *Matematičeskij analiz. Vtoroj specialnyj kurs*, Izd. MGU, Moskva, 1984.

[Sp]   M. Spivak, *Calculus*, Benjamin, London, 1973.

[Te]   S. A. Teliakovskij, *Sbornik zadač po teorii funkcij dejstvitelnogo peremennogo*, Nauka, Moskva, 1980.

[Th]   J. A. Thorpe, *Elementary Topics in Differential Geometry*, Springer, New York, 1979.

[VLA]   L. I. Volkovyskij, L. Lunc, I. G. Aramanovič, *Sbornik zadač po teorii funkcij kompleksnogo peremennogo*, Nauka, Moskva, 1975. Translation: *A Collection of Problems on Complex Analysis*, Pergamon Press, Oxford, 1965.

[W]   A. Wilansky, *Functional Analysis*, Blaisdell, New York, 1964.

[Zu]   C. Zuily, *Problèmes de distributions avec solutions détaillées*, Hermann, Paris, 1986. Translation: *Problems in Distributions and Partial Differential Equations*, North Holland, Amsterdam, 1988.

Index

Following each entry is the problem number in which the entry may be found.

Ingram Content Group UK Ltd.
Milton Keynes UK
UKHW020157180323
418782UK00013B/75